Communications
in Computer and Information Science 648

Commenced Publication in 2007
Founding and Former Series Editors:
Alfredo Cuzzocrea, Dominik Ślęzak, and Xiaokang Yang

More information about this series at http://www.springer.com/series/7899

Günter Fahrnberger · Gerald Eichler
Christian Erfurth (Eds.)

Innovations for Community Services

16th International Conference, I4CS 2016
Vienna, Austria, June 27–29, 2016
Revised Selected Papers

 Springer

Editors
Günter Fahrnberger
University of Hagen
Hagen
Germany

Gerald Eichler
Telekom Innovation Laboratories
Deutsche Telekom AG
Darmstadt, Hessen
Germany

Christian Erfurth
Ernst Abbe University of Applied Sciences
Jena
Jena
Germany

ISSN 1865-0929   ISSN 1865-0937 (electronic)
Communications in Computer and Information Science
ISBN 978-3-319-49465-4   ISBN 978-3-319-49466-1 (eBook)
DOI 10.1007/978-3-319-49466-1

Library of Congress Control Number: 2016957377

Printed on acid-free paper

This Springer imprint is published by Springer Nature
The registered company is Springer International Publishing AG
The registered company address is: Gewerbestrasse 11, 6330 Cham, Switzerland

Foreword

Since 2014, under its revised name, the International Conference on Innovations for Community Services (I4CS) has continued its success story. It developed from the national German IICS workshop, founded in 2001, toward a small but remarkable international event.

Traditionally, the conference alternates between international and German locations, which the Steering Committee members select year by year. As usual, the annual three-day event took place around the second half of June. Its name is its mission: to form an innovative community comprising scientists, researchers, service providers, and vendors. In 2016, we went to Austria for the first time.

After a two-year operation with IEEE, the I4CS Steering Committee decided to publish the proceedings with Springer CCIS as a new partner, to allow also leading technology companies, e.g., in telecommunications to be a so-called financial sponsor, acting as host and organizer for the conference.

The present proceedings comprise six session topics plus two short papers, covering the selection of the best papers from 2016 out of 30 submissions.

The scope of I4CS topics for 2016 spanned a unique choice of aspects, bundled into the three areas: "technology," "applications," and "socialization." Big data analytics had a strong focus in this year.

Technology – Distributed Architectures and Frameworks

- Infrastructure and models for community services
- Data structures and management in community systems
- Community self-organization in ad-hoc environments
- Search, information retrieval, and distributed ontology
- Smart world models and big data analytics

Applications – Communities on the Move

- Social networks and open collaboration
- Social and business aspects of user-generated content
- Recommender solutions and expert profiles
- Context and location awareness
- Browser application and smartphone app implementation

Socialization – Ambient Work and Living

- eHealth challenges and ambient-assisted living
- Intelligent transport systems and connected vehicles
- Smart energy and home control
- Social gaming and cyber physical systems
- Security, identity, and privacy protection

Many thanks to both the members of the Program Committee and all the volunteers for their flexible support during the preparation phase of I4CS 2016 and the host T-Mobile Austria, namely, Günter Fahrnberger, whose tremendous personal effort moved the tradition forward.

Besides the *Best Paper Award*, based on the ratings of the 25 members of the technical Program Committee, and the *Best Presentation Award*, chosen by all conference participants, the 2014 newly introduced *Young Scientist Award* was given for the second time.

The 17th I4CS conference, hosted by the Telekom Innovation Laboratories, will take place in Darmstadt, Germany, during June 21–23, 2017. Please check the conference website at http://www.i4cs-conference.org/ regularly for more details. Any new ideas and proposals for the future of the I4CS are welcome (request@i4cs-conference.org).

June 2016   Gerald Eichler

Preface

This book contains the papers presented at I4CS 2016, the 16th International Conference on Innovations for Community Services, held during June 27–29, 2016, in Vienna.

There were 30 submissions. Each submission was reviewed by at least two Program Committee members. The committee decided to accept 12 full papers and two short papers. This volume also includes the three invited talks.

I would like to dedicate this preface to all the contributors who made I4CS 2016 successful. Apart from the Program Committee, they were:

- Franziska Bauer and Stefanie Leschnik from T-Mobile Austria Corporate Communications
- Timm Herold from T-Systems Austria
- Walter Langer from T-Mobile Network Operations

June 2016   Günter Fahrnberger

Organization

Program Committee

Marwane Ayaida	University of Reims, France
Gilbert Babin	HEC Montréal, Canada
Martin Ebner	Graz University of Technology, Austria
Gerald Eichler	Telekom Innovation Laboratories, Germany
Christian Erfurth	EAH Jena, Germany
Günter Fahrnberger	University of Hagen, North Rhine-Westphalia, Germany
Hacene Fouchal	Université de Reims Champagne-Ardenne, France
Peter Kropf	Université de Neuchâtel, Switzerland
Ulrike Lechner	Bundeswehr University Munich, Germany
Karl-Heinz Lüke	Ostfalia University of Applied Sciences, Germany
Venkata Swamy Martha	WalmartLabs, USA
Phayung Meesad	King Mongkut's University of Technology North Bangkok, Thailand
Hrushikesha Mohanty	University of Hyderabad, India
Raja Natarajan	Tata Institute of Fundamental Research, India
Prasant K. Pattnaik	KIIT Universtiy, India
Chittaranjan Pradhan	KIIT Universtiy, India
Davy Preuveneers	University of Leuven, Belgium
Srini Ramaswamy	ABB Inc., USA
Wilhelm Rossak	Friedrich Schiller University Jena, Germany
Jörg Roth	Nuremberg Institute of Technology, Germany
Volkmar Schau	Friedrich Schiller University Jena, Germany
Julian Szymanski	Gdansk University of Technology, Poland
Martin Werner	Ludwig Maximilian University of Munich, Germany
Leendert W.M. Wienhofen	SINTEF, Norway
Ouadoudi Zytoune	Ibn Tofail University, Morocco

Additional Reviewers

Atluri, Vani Vathsala
Geyer, Frank
Lakshmi, H.N.
Vaddi, Supriya

Abstracts of Invited Talks

Enriching Community Services: Making the Invisible 'P' Visible

Srini Ramaswamy

ABB Incorporation, Cleveland, Ohio, USA
srini@ieee.org

Abstract. This talk will focus on embracing technological disruptions while simultaneously delivering meaningful community services. Much of our current day problems with large scale systems can be attributed to the inherent flexibility that users' actively seek in software-driven systems. Often, problems arise as these systems are not effectively designed and tested to coexist with other complex systems, including humans, who are vast and dynamic information elements within the systems' operational environment. This talk with take a multi-stakeholder perspective to the design of community service applications and zero-in on prioritizations that bring together these different stakeholder perspectives for delivering meaningful user experiences. Critical issues include assembling, integrating and analyzing information from disparate sources in a timely, accurate and reliable manner, while meeting real-time needs and expectations.

The Relevance of Off-the-Shelf Trojans in Targeted Attacks

Recent Developments and Future Obstacles in Defense

Marion Marschalek

G DATA Advanced Analytics, Bochum, North Rhine-Westphalia, Germany
marion@0x1338.at

Abstract. The malware landscape has changed drastically since the times when the term was first coined. As systems are becoming more complex the threats turn less and less comprehensible, naturally.

Following an introduction to the nature of targeted attacks, the audience will learn how modern day threat detection works and occasionally fails in regard to detecting malicious software. For more than three decades protection systems have relied on pattern recognition, and are now facing threats bare of any obvious patterns. As a case study a closer look at the fairly well documented compromise of Hacking Team's network will be taken, and correlated with current APT detection technologies.

On the contrary to highly sophisticated attacks, another trend we see in digital espionage is the heavy use of so called off-the-shelf RATs; ready made Remote Access Trojans, dedicated to carry out one of the final steps of a spy campaign - the data collection. As recent analysis shows, off-the-shelf RATs make up nearly a quarter of malicious binaries leveraged in targeted attacks.

In comparing a considerably advanced attack with a targeted attack heavily relying on reuse of tools and techniques one can introduce metrics regarding evasiveness and stealth, but also the so far rather neglected metric of costs of attack. While advances in awareness and operating system security tend to make offense more expensive, the use of ready made attack components drives down cost of development and maintenance. This session will provide an overview of state-of-the-art attack tools and techniques, threat detection measures, their applicabilities and weaknesses. Additionally the attack cost metric will be taken into account in the light of supporting protection mechanisms.

When Learning Analytics Meets MOOCs - a Review on iMooX Case Studies

Mohammad Khalil and Martin Ebner

Educational Technology, Graz University of Technology, Graz, Austria
{Mohammad.khalil,martin.ebner}@tugraz.at

Abstract. The field of Learning Analytics has proven to provide various solutions to online educational environments. Massive Open Online Courses (MOOCs) are considered as one of the most emerging online environments. Its substantial growth attracts researchers from the analytics field to examine the rich repositories of data they provide. The present paper contributes with a brief literature review in both prominent fields. Further, the authors overview their developed Learning Analytics application and show the potential of Learning Analytics in tracking students of MOOCs using empirical data from iMooX.

Contents

Invited Talk

When Learning Analytics Meets MOOCs - a Review on iMooX
Case Studies. 3
   Mohammad Khalil and Martin Ebner

Navigation and Data Management

The Offline Map Matching Problem and its Efficient Solution 23
   Jörg Roth

Using Data as Observers: A New Paradigm for Prototypes Selection 39
   Michel Herbin, Didier Gillard, and Laurent Hussenet

Monitoring and Decision Making

Reconstruct Underground Infrastructure Networks Based on
Uncertain Information 49
   Marco de Koning and Frank Phillipson

Design and Realization of Mobile Environmental Inspection and
Monitoring Support System 59
   Hyung-Jin Jeon, Seoung-Woo Son, Jeong-Ho Yoon, and Joo-Hyuk Park

Coding and Security

Re-visited: On the Value of Purely Software-Based Code Attestation for
Embedded Devices 75
   Maximilian Zeiser and Dirk Westhoff

Secure Whitelisting of Instant Messages. 90
   Günter Fahrnberger

Collaboration and Workflow

Adaptive Workflow System Concept for Scientific Project Collaboration 115
   Vasilii Ganishev, Olga Fengler, and Wolfgang Fengler

Zebras and Lions: Better Incident Handling Through Improved Cooperation . . . 129
   Martin Gilje Jaatun, Maria Bartnes, and Inger Anne Tøndel

Routing and Technology

Routing over VANET in Urban Environments   143
  Boubakeur Moussaoui, Salah Merniz, Hacène Fouchal,
  and Marwane Ayaida

Tech4SocialChange: Technology for All   153
  André Reis, David Nunes, Hugo Aguiar, Hugo Dias, Ricardo Barbosa,
  Ashley Figueira, André Rodrigues, Soraya Sinche, Duarte Raposo,
  Vasco Pereira, Jorge Sá Silva, Fernando Boavida, Carlos Herrera,
  and Carlos Egas

Topic and Object Tracking

Topic Tracking in News Streams Using Latent Factor Models   173
  Jens Meiners and Andreas Lommatzsch

Collaboration Support for Transport in the Retail Supply Chain:
A User-Centered Design Study   192
  Marit K. Natvig and Leendert W.M. Wienhofen

Short Papers

Potentials and Requirements of an Integrated Solution for a Connected Car   211
  Karl-Heinz Lüke, Gerald Eichler, and Christian Erfurth

ICT-Systems for Electric Vehicles Within Simulated and Community
Based Environments   217
  Volkmar Schau, Sebastian Apel, Kai Gebhard, Marianne Mauch,
  and Wilhelm Rossak

Author Index   223

Invited Talk

When Learning Analytics Meets MOOCs - a Review on iMooX Case Studies

Mohammad Khalil[⊠] and Martin Ebner

Educational Technology, Graz University of Technology, Graz, Austria
{Mohammad.khalil,martin.ebner}@tugraz.at

Abstract. The field of Learning Analytics has proven to provide various solutions to online educational environments. Massive Open Online Courses (MOOCs) are considered as one of the most emerging online environments. Its substantial growth attracts researchers from the analytics field to examine the rich repositories of data they provide. The present paper contributes with a brief literature review in both prominent fields. Further, the authors overview their developed Learning Analytics application and show the potential of Learning Analytics in tracking students of MOOCs using empirical data from iMooX.

Keywords: Learning analytics · Massive open online courses (MOOCs) · Completionrate · Literature · Engagement · Evaluation · Prototype

1 Introduction

The growth of Massive Open Online Courses (MOOCs) in the modernistic era of online learning has seen millions of enrollments from all over the world. They are defined as online courses that are open to the public, with open registration option and open-ended outcomes that require no prerequisites or fees [23]. These courses have brought a drastic action to the Higher Education from one side and to the elementary education from the other side [13]. The number of offered MOOCs has exploded in the recent years. Particularly, until January 2016, there have been over 4500 courses with 35 million learners from 12 MOOC providers [25]. Some of these courses are provided by prestigious and renowned universities such as Harvard, MIT, and Stanford. At the same time, other institutions have joined the MOOC hype and became providers of their own local universities like the Austrian MOOC platform, iMooX (www.imoox.at).

It is important to realize that MOOCs have split into two major types: cMOOCs and xMOOCs. The cMOOCs are based on the philosophy of connectivism which is about creating networks of learning [27]. On the other hand, the xMOOCs term is shortened from extended MOOCs based on classical information transmission [10]. Further, new types of online courses related to MOOCs have germinated recently such as Small Private Online Courses (SPOCs) and Distributed Open Collaborative Courses (DOCCs).

MOOCs have the potential of scaling education in different fields and subjects. The study of [25] showed that computer science and programming grabbed the largest percentage of the offered courses. Yet, substantial growth of MOOCs has also been

© Springer International Publishing AG 2016
G. Fahrnberger et al. (Eds.): I4CS 2016, CCIS 648, pp. 3–19, 2016.
DOI: 10.1007/978-3-319-49466-1_1

noticed in Science, Technology, Engineering, and Mathematics (STEM) fields. The anticipated results of MOOCs were varied between business purposes like saving costs, and improving the pedagogical and educational concepts of online learning [16]. Nevertheless, there is still altercation about the pedagogical approach of information delivery to the students. The quality of the offered courses, completion rate, lack of interaction, and grouping students in MOOCs have been, in addition, debated recently [4, 12, 17].

Since MOOCs are an environment of online learning, the educational process is based on video lecturing. In fact, learning in MOOCs is not only exclusive to that, but social networking and active engagement are major factors too [23]. Contexts that include topics, articles or documents are also considered as a supporting material in the learning process.

While MOOC providers initialize and host online courses, the hidden part embodied in recording learners' activities. Nowadays, ubiquitous technologies have spread among online learning environments and tracking students online becomes much easier. The pressing needs of ensuring that the audience of eLearning platforms is getting the most out of the online learning process and the needs to study their behavior lead to what is so-called "Learning Analytics". One of its key aspects is identifying trends, discovering patterns and evaluating learning environments, MOOCs here as an example. Khalil and Ebner listed factors that have driven the expansion of this emerging field [14]: (A) technology spread among educational categories, (b) the "big data" available from learning environments, and (c) the availability of analytical tools.

In this research publication, we will discuss the potential of the collaboration between Learning Analytics and MOOCs. There have been various discussions among researchers from different disciplines regarding these apparent trends. For instance, Knox said that "Learning Analytics promises a technological fix to the long-standing problems of education" [19]. Respectively, we will line up our experience within both of the fields in the recent years and list the up to date related work. Further, different scenarios and analysis from offered MOOCs of the iMooX will be discussed using the iMooX Learning Analytics Prototype. At the end, we will list our proposed interventions that will be adopted in the next MOOCs.

This publication is organized as follows: Sect. 2 covers literature and related work. In Sect. 3, we list a systematic mapping from the Scopus library to understand what has been researched in Learning Analytics of MOOCs. Section 4 covers the iMooX Learning Analytics Prototype while Sect. 5 covers case studies and the derived analytics outcomes from the empirically provided data.

2   Literature Review

2.1   MOOCs

The new technologies of the World Wide Web, mobile development, social networks and the Internet of Things have advanced the traditional learning. eLearning and Technology Enhanced Learning (TEL) have risen up with new models of learning environments such as Personal Learning Environments (PLE), Virtual Learning

Environments (VLE) and MOOCs. Since 2008, MOOCs reserved a valuable position in educational practices. Non-profits platforms like edX (www.edx.org) and profit platforms like Coursera (www.coursera.com) attracted millions of students. As long as they only require an Internet connection and intention for learning, MOOCs are considered to be welfare for the Open Educational Resources (OER) and the lifelong learning orientation [7].

Despite all these benefits, MOOCs turn out badly with several issues. Dropout and the failure to complete courses are considered as one of the biggest issues. Katy Jordan showed that the completion rate of many courses merely reached 10 % [11]. Reasons behind were explained because of poor course design, out of motivation, course takes much time, lack of interaction and the assumption of too much knowledge needed [16, 21]. Fetching other issues of MOOCs through available empirical data is discussed later in this paper.

2.2 Learning Analytics

The birth of Learning Analytics has first seen the light in 2011. A Plethora of definitions were used since then. However, the trend is strongly associated with previously well-known topics such as web analytics, academic analytics, data analysis, data mining as well as psychometrics and educational measurement [2]. Learning Analytics mainly targets educational data sets from the modern online learning environments where learners leave traces behind. The process then includes searching, filtering, mining and visualizing data in order to retrieve meaningful information.

Learning Analytics involves different key methods of analysis. They vary from data mining, statistics, and mathematics, text analysis, visualizations, social network analysis, qualitative to gamification techniques [15, 26]. On the other hands, the aims of Learning Analytics diversify between different frameworks, but most of them agreed on common goals. Despite its learning environment, Papamitsiou and Economides showed that studies of Learning Analytics focused on the pedagogical analysis of behavior modeling, performance prediction, participation and satisfaction [26]. Benefits utilized in prediction, intervention, recommendation, personalization, evaluation, reflection, monitoring and assessment improvement [3, 9, 14]. In fact, these goals are considered useless without optimizing, refining and taking the full power of it on stakeholders [5].

2.3 MOOCs and Learning Analytics

Learners of the online learning environments such as MOOCs are not only considered as consumers, but they are also generators of data [14]. Lately, the research part of studying the behavior of online students in MOOCs becomes widely spread across journals and conferences. A recent survey study done by Khalil and Ebner on Learning Analytics showed that the ultimate number of citations using Google scholar (scholar.google.com) were relevant to MOOC articles [15]. They listed the most common techniques used by Learning Analytics in MOOCs, varying from machine learning, statistics, information visualization, Natural Language Processing (NLP), social network analysis, to gamification tools. Moissa and her colleagues mentioned that Learning Analytics in MOOCs literature studies are still not deeply researched [24]. We also found that valid in the next section.

3   Learning Analytics of MOOCs

In this section, we did a brief text analysis and mapped the screening of the abstracts from the Scopus database (www.scopus.com), in order to:

1. Grasp what has been researched in Learning Analytics of MOOCs.
2. Realize the main research trends of the current literature of Learning Analytics and MOOCs.

Scopus is a database powered by Elsevier Science. Our selection of this library is because of the valuable indexing information it provides and the usability of performing search queries. The conducted literature exploration was performed by searching for the following keywords: "Learning Analytics" and "MOOC", "MOOCs" or "Massive Open Online Course". The used query to retrieve the results was executed on 11- April- 2016 and is shown in Fig. 1. The language was refined to English only.

```
Your query : ((TITLE-ABS-KEY("Learning Analytics" AND "MOOCs") OR TITLE-ABS-
KEY("Learning Analytics" AND "MOOC") OR TITLE-ABS-KEY("Learning Analytics" AND
"Massive Open Online Course")) AND ( LIMIT-TO(LANGUAGE,"English" ) ) )
```

Fig. 1.   Search query to conduct the literature mapping

The returned results equaled to 80 papers. Only one paper was retrieved in 2011, none from 2012, 11 papers from 2013, 23 papers from 2014, 37 from 2015 and 8 papers from 2016. Abstracts were then extracted and processed to a Comma-Separated Values (CSV) file. After that, we created a word cloud in furtherance of representing text data to identify the most prominent terms. Figure 2 depicts the word cloud of the extracted abstracts. We looked at the single, bi-grams, tri-grams and quad-grams common terms. The most repeated single words were "MOOCs", "education", and "engagement". On the other hand, "Learning Analytics", "Online Courses" and "Higher Education" were recorded as the prominent bi-grams. "Khan Academy platform" and "Massive Open Online Courses" were listed on the top of the tri-grams and quad-grams respectively. As long as massive open online courses are represented in different terms in the abstracts, we abbreviated all the terms to "MOOCs" in the corpus.

Figure 3 shows the most frequent phrases fetched from the text. Figures 2 and 3 show interesting observations of the researched topics of Learning Analytics in MOOCs. By doing a simple grouping of the topics and disregarding the main phrases which are "Learning Analytics" and "MOOCs", we found that researchers were looking mostly at the engagement and interactions.

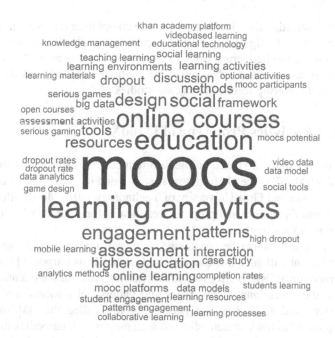

Fig. 2. Word cloud of the most prominent terms from the abstracts

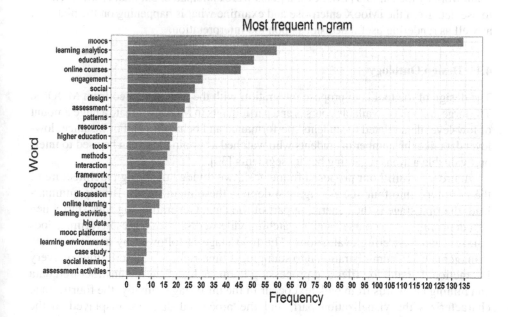

Fig. 3. The most frequent terms extracted from the abstracts

It was quite interesting that the dropout and the completion rate were not the major topics as we believed. Design and framework principles as well as assessment were ranked the second most cited terms. Social factors and learning as well as discussions grabbed the afterward attention, while tools and methods were mentioned to show the mechanism done in offering solutions and case studies.

4   Learning Analytics Prototype of iMooX

The analyses of this study are based on the different courses provided by the Austrian MOOC provider (iMooX). The platform was first initiated in 2013 with the cooperation of University of Graz and Graz University of Technology [20]. iMooX offers German courses in different disciplines and proposes certificates for students who successfully complete the courses for free.

A MOOC platform cannot be considered as a real modern technology enhanced learning environment without a tracking approach for analysis purposes [16]. Tracking students left traces on MOOC platforms with a Learning Analytics application is essential to enhance the educational environment and understand students' needs. iMooX pursued the steps and applied an analytical approach called the "iMoox Learning Analytics Approach" to track students for research purposes. It embodies the functionality to interpret low-level data and present them to the administrators and researchers. The tool is built based on the architecture of the early presented Learning Analytics framework by the authors [14]. Several goals were anticipated, but mainly the intention to use data from the iMooX enterprise and examine what is happening on the platform as well as rendering useful decisions upon the interpretation.

4.1   Design Ontology

The design of the tool is to propose integration with the data generated from MOOCs. The large amount of available courses and participants in MOOCs, create a huge amount of low-level data related to students' performance and behavior [1]. For instance, low-level data like the number of students who watched a certain video can be used to interpret valuable actions regarding boring segments [30].

In order to fulfill our proposed framework, we divided the design architecture of the prototype into four stages. Figure 4 depicts these main stages. Briefly summarized, the first stage is the generation part of the data. Generating log files start when a student enrolls in a course, begins watching videos, discusses topics in forums, does quizzes, and answering evaluations. The next stage is followed by a suitable data management and administration into stamping a time-reference descriptions of every interaction. Parsing log files and processing them such as filtering unstructured data and mining keywords from bulk text occur in the third stage. Finally, the fourth stage characterizes the visualization part, and the processed data are displayed to the admins and researchers.

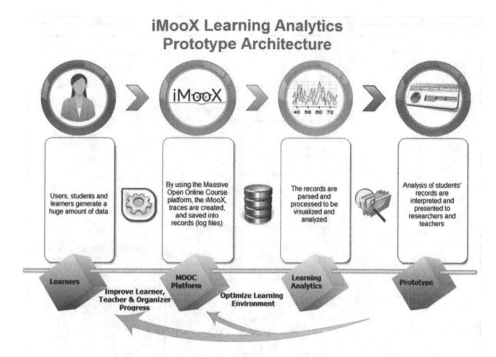

Fig. 4. The iMooX Learning Analytics Prototype design architecture [16]

4.2 Implementation Architecture and User Interface

The implementation framework adopts the design architecture with more detailed processing steps for the visualization part. We aimed to develop an easy-to-read dashboard. The intended plan was to make visualizations for taking actions. They should not only be connected with meaning and facts [6]. Thus, the data are presented in a statistical text format and in charts like pie charts and bar plots as shown below in Fig. 5.

This user dashboard is only accessible by researchers and administrators. A teacher version, however, is attainable in a static format which shows general statistics about his/her teaching course. The detailed personal information of students is kept confidential and is only available for research and administrative reasons. The Dashboard shows various MOOC objects and indicators. These objects inherent pedagogical purposes and require appropriate interpretation for proper actions [8]. The Dashboard offers searching for any specific user in a particular period. The returned results cover:

- Quiz attempt, scores, and self-assessment
- Downloaded documents from the course
- Login frequency
- Forums reading frequency
- Forums posting frequency
- Watched videos

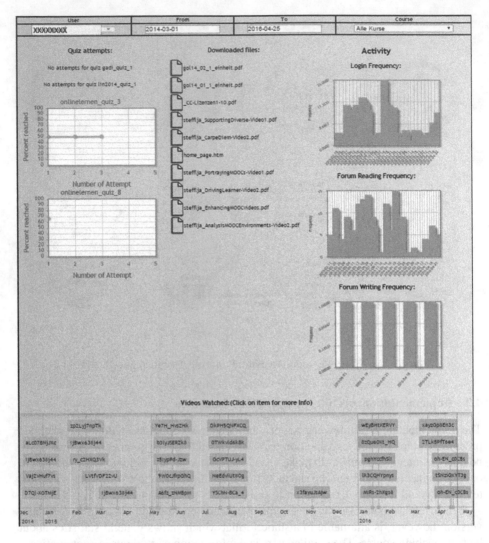

Fig. 5. iMooX Learning Analytics Prototype user dashboard - admin view

Further, comprehensive details can be carried out of each indicator when required by clicking on the learning object tab.

5   Analysis and Case Studies

This section shows some of detailed analyses done previously. This examination is carried out using the log data fetched from the prototype. The awaited results are: (a) evaluating the prototype efficiency in revealing patterns, (b) recognizing the potentiality of Learning Analytics in MOOCs.

5.1 Building Activity Profiles

Building an activity profile using Learning Analytics becomes possible using the rich available data provided by the prototype. We have analyzed a MOOC called "Mechanics in Everyday life". The course was ten weeks long, and the target group was secondary school students from Austria. The MOOC, however, was also open to the public. There were (N = 269) participants. The aim behind the activity profile is to deeply examine the activity of participants and to distinguish between their activities. Figure 6 displays the activity profile only for school pupils. The green represents the certified students (N = 5), while the red represents the non-certified students (N = 27). It is obvious that week-1, week-3, and week-4 were very active in discussion forums. Watching videos were totally uninteresting in the last week. Thorough observations and differences between pupils and other enrollees can be trailed from [13].

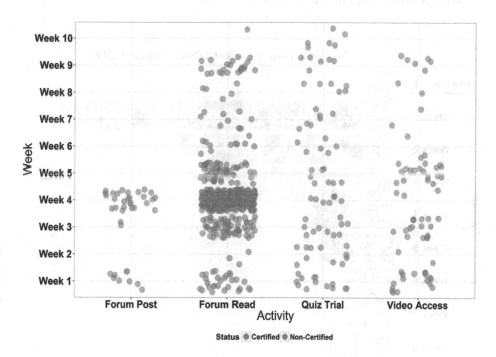

Fig. 6. The activity profile

5.2 Tracking Forums Activity

Various discussions about the role of social activity in MOOCs forums were regularly debated. Recently, the study by Tseng et al., found out that the activity in forum discussion is strongly related to the course retention and performance [28]. We have done several exploratory analyses to uncover diverse pedagogical relations and results

[16, 17, 21, 22]. The analyses are based on different offered MOOCs. The following outcomes were concluded:

- Defined drop-out point where students posting and reading in forums clearly diminishes in week-4, as shown in Fig. 7. We found such patterns being recurred among a collection of MOOCs.
- Figure 8 shows the relation between reading and writing in the discussion forums. Different samples were tested randomly. The Pearson-moment correlation coefficient of 0.52 and p-value < 0.01 was calculated. This indicates a moderate correlation of students who write more are likely to read more. Further, the active instructor drives a positive interaction into creating a dynamic social environment.
- Figure 9 depicts forum posts in two courses. Students usually write more often in the first two weeks.
- Figure 10 shows the timing trends of learning happening during the whole day. Peaks were detected between 6 p.m. and 10 p.m.

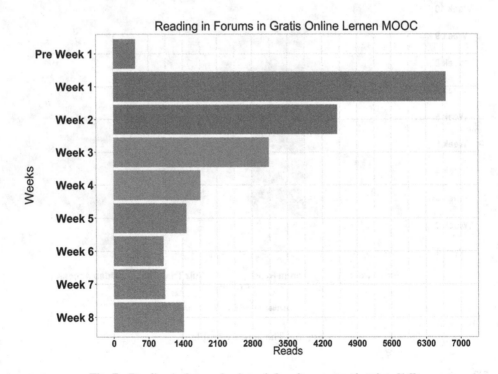

Fig. 7.   Reading in forums leads to define drop-out peak points [16]

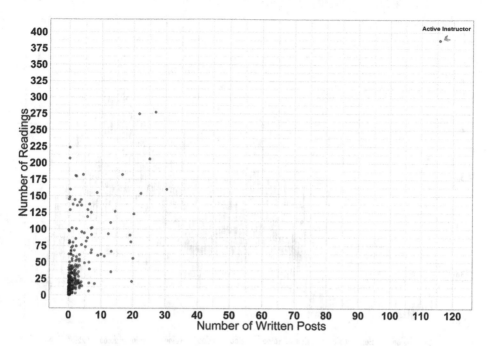

Fig. 8. Positive relationship between reading and writing in forums [22]

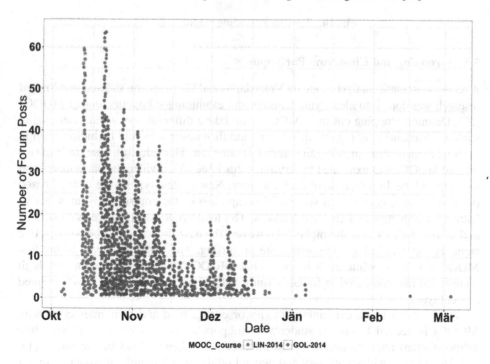

Fig. 9. Participants discuss and ask more often in the first two weeks [16]

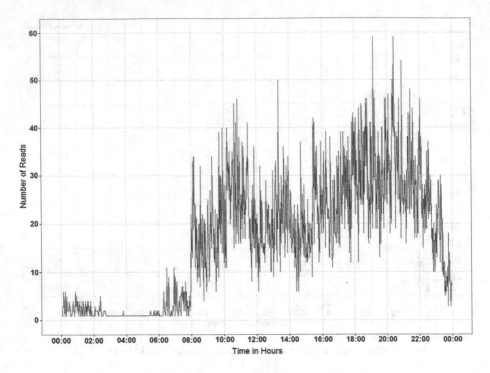

Fig. 10. Time spent reading in forums [22]

5.3 Grouping and Clustering Participants

A recent systematic analysis done by Veletsianos and Shepherdson showed that limited research was done into identifying learners and examining subpopulations of MOOCs [29]. Defining dropping out from MOOCs can take a different way when categorizing students. Students may register in a course and then never show up. Including them in the total dropout share implies unjustified retention rate. Henceforth, a case study of two offered MOOCs was examined to scrutinize this issue. We divided the participants and investigated the dropout ratio of each category. New student types were defined based on their activity, quizzes and successful completion of the course: registrants, active learners, completers and certified learners. The total dropout gap between registrants and active students was the highest. However, the new dropout ratio between active students and certified learners was quite promising. The completion rate in the first MOOC was 37 %, while 30 % in the second MOOC. This is considered a very high completion rate compared to Jordan's study [11]. Figure 11 shows the newly defined student types.

On the other hand, explaining activity or engagement and their interaction with MOOCs is needed to cluster students to subpopulations. Classifying students into subpopulations improve decisions and interventions taken by lead managements [17, 18]. However, engagements vary and depend on the tested sample of students. In our paper "Portraying MOOCs learners: a clustering experience using learning analytics",

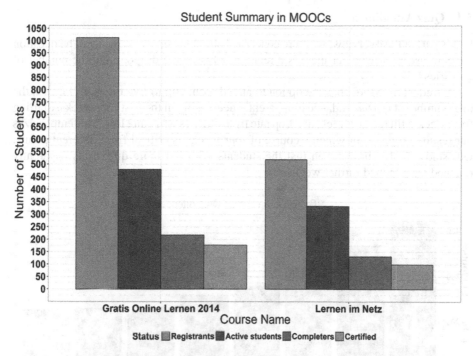

Fig. 11. New types of MOOC learners are defined using Learning Analytics [16]

we studied the engagement of university students using the k-means clustering algorithm [17]. Table 1, shows the activities of each cluster (reading frequency, writing frequency, watching videos, quiz attendance, and certification ratio).

Table 1. University students clustering results [17]

Cluster	Read F.	Write F.	Watch Vid.	Quiz Att.	Cert. Ratio
C1	Low	Low	Low	Low	10.53 %
C2	High	Low	High	High	96.10 %
C3	Moderate	Low	Low	High	94.36 %
C4	High	High	Low	Moderate	50 %

Four sorts of students were detected using the algorithm. The "dropout" cluster shown as C1 is clarified by students who have a low activity in all MOOC learning activities. "On track" or excellent students displayed as C2 in the table and those who are involved in most of the MOOC activities and have a certification rate of 96.1 %. The "Gamblers" or students who play the system shown as C3, and these barely watch learning videos, but they did every quiz seeking for the grade. "Social" students, shown as C4 and these are more engaged in forums and their certification rate was around 50 %.

5.4 Quiz Attendance

In this part of the overview, we were concerned about the quiz attendance. The question seeks to ascertain whether there is a relation between the dropout and the number of quiz tries?

A student in iMooX has the option to attend a quiz up to five times. In Fig. 12, the total number of quiz attendance is apparently decreasing in the first four weeks. Starting from week-5 till the last week, the drop rate from quizzes was quite low. This emphasizes our results in Fig. 7, in which a course of four weeks is critical from different points. Our study in [13] has proven that the students who did more quiz trials apparently retained and reached further weeks.

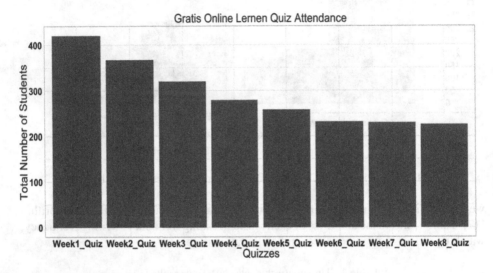

Fig. 12. Quiz attendance in one of the MOOCs (eight weeks long)

6 Discussion and Conclusion

This research study is mainly divided into three parts. The first part lists literature study of the topics of Learning Analytics, MOOCs, and Learning Analytics in MOOCs. With the 80 collected papers from the Elsevier Science library: Scopus, we did a word cloud to remark the vital trends in these two prominent fields. Topics of engagement, interactions, social factors, as well as design and frameworks, were referred the most. Further, Learning Analytics was employed more into improving interactions and engagements of students in the MOOCs environment instead of the dropout matter. The second part shows our experience in implementing the iMooX Learning Analytics prototype. It eases collecting and tracking students of the examined MOOC platform. We discussed its ontology, implementation architecture and user interface. The third part evaluated the application. Different scenarios from iMooX were analyzed using advanced visualizations, statistics, clustering and qualitative decisions.

The potentiality of Learning Analytics in MOOCs crystallizes in the subsequent interventions upon the evaluation results. We believe in designing shorter courses such as four weeks MOOC instead of eight weeks [21]. As a result, the workload would be cut in half, and students' efficiency will be higher. Additionally, the concept of enhancing social communications in the discussion forums, especially between the instructor and the students would attract students into being connected which by all means would decrease the dropout rate. We further discovered new types of students using categorization and clustering by depending on their activity. This will lead us into portraying engagement and behavior of a subpopulation of learners in the platform.

We think learning analytics carries significant values to MOOCs from the pedagogical and technological perspectives. Proper interventions, predictions, and benchmarking learning environments are difficult to optimize on MOOCs without the assistance of Learning Analytics. In the end, the under development algorithm of designing an assistant tool that sends a direct feedback to the student in order to improve the completion rate is in our future plans. It will notify students directly in order to support a live awareness and reflection system.

References

1. Alario-Hoyos, C., Muñoz-Merino, P.J., Pérez-Sanagustín, M., Delgado Kloos, C., Parada, G.H.A.: Who are the top contributors in a MOOC? Relating participants' performance and contributions. J. Comput. Assist. Learn. **32**(3), 232–243 (2016)
2. Baker, R.S., Siemens, G.: Educational Data Mining and Learning Analytics (2016).http://www.columbia.edu/~rsb2162/BakerSiemensHandbook2013.pdf. Accessed 20 Apr 2016
3. Chatti, M., Dyckhoff, A., Schroeder, U., Thüs, H.: A reference model for learning analytics. Int. J. Technol. Enhanced Learn. **4**(5/6), 318–331 (2012)
4. Clow, D.: MOOCs and the funnel of participation. In: The Third International Conference on Learning Analytics and Knowledge (LAK 2013), Leuven, Belgium, pp. 185–189. ACM (2013)
5. Clow, D.: The learning analytics cycle: closing the loop effectively. In: The 2nd International Conference on Learning Analytics and Knowledge (LAK 2012), Vancouver, Canada, pp. 134–138. ACM (2012)
6. Duval, E.: Attention please!: Learning analytics for visualization and recommendation. In: The 1st International Conference on Learning Analytics and Knowledge (LAK 11), Alberta, Canada, pp. 9–17. ACM (2011)
7. Ebner, M., Schön, S., Kumar, S.: Guidelines for leveraging university didactics centers to support OER uptake in German-speaking Europe. Educ. Policy Anal. Arch. **24**, 39 (2016)
8. Graf, S., Ives, C., Rahman, N., Ferri, A.: AAT: a tool for accessing and analysing students' behaviour data in learning systems. In: The 1st International Conference on Learning Analytics and Knowledge (LAK 2011), Alberta, Canada, pp. 174–179. ACM (2011)
9. Greller, W., Drachsler, H.: Translating learning into numbers: a generic framework for learning analytics. Educ. Technol. Soc. **15**(3), 42–57 (2012)
10. Hollands, F.M., Tirthali, D.: MOOCs: Expectations and reality. Full report. Center for Benefit-Cost Studies of Education, Teachers College, Columbia University, NY (2014). http://cbcse.org/wordpress/wp-content/uploads/2014/05/MOOCs_Expectations_and_Reality.pdf. Accessed 19 April 2016

11. Jordan, K.: MOOC completion rates: the data (2013). http://www.katyjordan.com/MOOC project.html. Accessed: 12 April 2016
12. Khalil, H., Ebner, M.: MOOCs completion rates and possible methods to improve retention - a literature review. In: Proceedings of World Conference on Educational Multimedia, Hypermedia and Telecommunications 2014, pp. 1236–1244. Chesapeake, VA: AACE (2014)
13. Khalil, M., Ebner, M.: A STEM MOOC for school children—what does learning analytics tell us? In: The 2015 International Conference on Interactive Collaborative Learning (ICL 2015), Florence, Italy, pp. 1217–1221. IEEE (2015)
14. Khalil, M., Ebner, M.: Learning analytics: principles and constraints. In: Carliner, S., Fulford, C., Ostashewski, N. (eds.) Proceedings of EdMedia: World Conference on Educational Media and Technology 2015, pp. 1789–1799. AACE, Chesapeake (2015)
15. Khalil, M., Ebner, M.: What is learning analytics about? A survey of different methods Used in 2013–2015. In: The Smart Learning Conference, Dubai, UAE, pp. 294–304. Dubai: HBMSU Publishing House (2016)
16. Khalil, M., Ebner, M.: What massive open online course (MOOC) stakeholders can learn from learning analytics? In: Spector, M., Lockee, B., Childress, M. (eds.) Learning, Design, and Technology: An International Compendium of Theory, Research, Practice, and Policy, pp. 1–30. Springer, Heildelberg (2016). http://dx.doi.org/10.1007/978-3-319-17727-4_3-1
17. Khalil, M., Kastl, C., Ebner, M.: Portraying MOOCs learners: a clustering experience using learning analytics. In: The European Stakeholder Summit on experiences and best practices in and around MOOCs (EMOOCS 2016), Graz, Austria, pp. 265–278 (2016)
18. Kizilcec, R.F., Piech, C., Schneider, E.: Deconstructing disengagement: analyzing learner subpopulations in massive open online courses. In: The Third International Conference on Learning Analytics and Knowledge (LAK 2013), Leuven, Belgium, pp. 170–179. ACM (2013)
19. Knox, J.: From MOOCs to learning analytics: scratching the surface of the 'visual'. eLearn. **2014**(11) (2014). ACM
20. Kopp, M., Ebner, M.: iMooX - Publikationen rund um das Pionierprojekt. Verlag Mayer, Weinitzen (2015)
21. Lackner, E., Ebner, M., Khalil, M.: MOOCs as granular systems: design patterns to foster participant activity. eLearning Pap. **42**, 28–37 (2015)
22. Lackner, E., Khalil, M., Ebner, M.: How to foster forum discussions within MOOCs: A case study. Int. J. Acad. Res. Educ. (in press)
23. McAulay, A., Tewart, B., Siemens, G.: The MOOC model for digital practice. Charlottetown: University of Prince Edward Island (2010). http://www.elearnspace.org/Articles/MOOC_Final.pdf
24. Moissa, B., Gasparini, I., Kemczinski, A.: A systematic mapping on the learning analytics field and its analysis in the massive open online courses context. Int. J. Distance Educ. Technol. **13**(3), 1–24 (2015)
25. Online Course Report.: State of the MOOC 2016: A Year of Massive Landscape Change For Massive Open Online Courses (2016). http://www.onlinecoursereport.com/state-of-the-mooc-2016-a-year-of-massive-landscape-change-for-massive-open-online-courses/. Accessed 18 April 2016
26. Papamitsiou, Z., Economides, A.A.: Learning analytics for smart learning environments: a meta-analysis of empirical research results from 2009 to 2015. In: Spector, M.J., Lockee, B.B., Childress, M.D. (eds.) Learning, Design, and Technology, pp. 1–23. Springer, Cham (2016)
27. Siemens, G.: A learning theory for the digital age. Instr. Technol. Distance Educ. **2**(1), 3–10 (2005)

28. Tseng, S.F., Tsao, Y.W., Yu, L.C., Chan, C.L., Lai, K.R.: Who will pass? Analyzing learner behaviors in MOOCs. Res. Pract. Technol. Enhanced Learn. **11**(1), 1–11 (2016)
29. Veletsianos, G., Shepherdson, P.: A systematic analysis and synthesis of the empirical MOOC literature published in 2013–2015. Int. Rev. Res. Open Distrib. Learn., **17**(2) (2016)
30. Wachtler, J., Khalil, M., Taraghi, B., Ebner, M.: On using learning analytics to track the activity of interactive MOOC videos. In: Proceedings of the LAK 2016 Workshop on Smart Environments and Analytics in Video-Based Learning, Edinburgh, Scotland, pp. 8–17. CEUR (2016)

Navigation and Data Management

The Offline Map Matching Problem
and its Efficient Solution

Jörg Roth[✉]

Faculty of Computer Science, Nuremberg Institute of Technology,
Nuremberg, Germany
Joerg.Roth@th-nuernberg.de

Abstract. In this paper we present an efficient solution to the offline map matching problem that occurs in the area of position measurements on a road network: given a set of measured positions, what was the most probable trip that led to these measurements? For this, we assume a user who moves according to a certain degree of optimality across a road network. A solution has to face certain problems; most important: as a single measurement may be mapped to multiple positions on the road network, the total number of combination exceeds any reasonable limit.

Keywords: Route planning · Optimal paths · Map matching · Geo data

1 Introduction

For many applications and services it is useful to know where a user was in the past. All positioning systems to track a user's position have certain measurements error. If we need to know a position very precisely, we have to introduce further mechanisms. Certain approaches are useful for movements on a road network: if a position is measured when driving on a road, we can perform a projection on the road network. This approach is called *map matching* and incorporated into common car navigation systems. As the mapping is performed immediately whenever a new measurement appears, this is a typical *online* approach. However, we can take further advantage of a road network, if we considered an entire trip afterwards: a reasonable projection on the road network should produce a route that could have been driven by the driver regarding her or his driving behaviour and degree of optimality. As we only can provide such a mapping after the trip was finished, we call this problem the *offline map matching problem*. A solution of this problem is useful for several applications:

- statistical analyses of drivers and flows of traffic;
- collecting street charges;
- automatically record a driver's logbook;
- learn typical trips from users, to e.g. automatically generate traffic warnings.

The research about map matching was historically strongly related to the *Global Positioning System* (GPS), where the major goal was to improve GPS measurements with help of a road network [13]. As a first mechanism, a position is mapped to the

© Springer International Publishing AG 2016
G. Fahrnberger et al. (Eds.): I4CS 2016, CCIS 648, pp. 23–38, 2016.
DOI: 10.1007/978-3-319-49466-1_2

geometrically nearest road position [11]. Further approaches took into account the road network's topology [2, 3, 9] also tries to address incomplete topologies. Based on the *multiple hypothesis technique* they follow multiple possible tracks that may approximate a sequence of measurements. An abstract approach to iterate through possible routes that approximate measurements was presented in [12], however, without solving the problem of the combinatorial explosion of route variations. [4] suggests a genetic algorithm to solve it.

Existing work relies on a high density of measurements (e.g. every 10 m), a generally high precision of the positioning system and measurement errors with nearly constant offset (of distance and direction) for longer time periods. The properties are fulfilled by satellite navigation systems. However, our approach in contrast should also work for sparse measurements (e.g. every 500 m) and for positioning systems based on signal strength fingerprinting currently used in smart phones. As a consequence, our approach has to fill gaps of unknown positions between measurements by a route planning approach. We strongly believe, a reasonable approach has to base on multi-routing capabilities to achieve a reasonable efficiency. I.e., we have to find optimal routes from multiple starts to multiple targets in a single call.

In this paper we often refer to the car driving scenario, however our approach is also suitable for other means of transportation (e.g. pedestrian, bicycle).

2 The Offline Map Matching Problem

2.1 Problem Statement

The basis of our considerations forms a *road network*: a topology of *crossings* and road segments between crossings, in the following called *links*. A routing function computes optimal paths (according to a cost function) between positions on the road network. In addition to the road topology, we need *link geometries* in order to map measurements to roads.

We can formulate the offline map matching as follows: given a sequence of measured positions that can be mapped to multiple possible positions on the road network; which combination of mapped positions represent the most probable driven route, if we connect all mapped positions by optimal paths?

Figure 1 illustrates the problem. We have a sequence of measurements denoted by 1–6. Figure 1a shows a simple map matching approach where we map each position to the nearest position that resides on the road network. If we now connect all mapped positions by optimal paths, we get a rather long path between 3 and 4. Moreover, we have U-turns at these positions. Figure 1b shows a more probable mapping: even though the mapping distances are bigger on average, the resulting path is more probable than in Fig. 1a as the overall route is shorter and avoids U-turns.

Note that we still talk about probabilities as we cannot definitely say, how a route exactly was. The first variation is possible for a driver, who, e.g. looked for a certain address, drove into a dead-end, turned and searched again. The lack of measurements that represent the long part between 3 and 4 could be explained by temporary measurement problems. However, we would consider Fig. 1b as a more appropriate

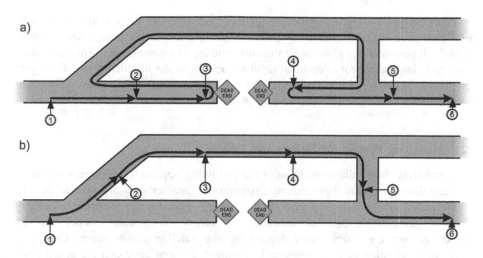

Fig. 1. General problem of mapping measurements to positions

assumption of the driven route, as it suits more to the measured positions and represents typical driving. Later we want to formalize 'typical' driving by models.

We first want to describe the offline map matching problem more formally. Let M be the set of possible measurements, P be the set of possible road positions and R be the set of routes on the road network for any pairs of start and target. Let $\wp(X)$ denote the set of *finite* subsets of X and $\mathscr{L}(X)$ the set of *finite* sequences of X. Further let

- *map*: $M \to \wp(P)$, $m_i \mapsto \{p_{i1} \ldots p_{ik}\}$ be the function that maps a measured position to a *finite* set of possible route positions;
- *route*: $P \times P \to R$, $(p_a, p_b) \mapsto r_{ab}$ be the function that maps two road positions to the optimal route to connect them in the given order;
- *eval*: $R \to \mathbb{R}$, $r \mapsto e_r$ be the function that evaluates a route and produces a value e_r that is lower for 'better' routes;
- $\mid$: $R \times R \to R$, $(r_a, r_b) \mapsto r_{ab}$ be the concatenation of routes that is only defined for routes r_a, r_b, where the first position of r_b is identical to the last position of r_a.

Then, we can formulate the (offline) map matching function as follows:

$$
\begin{aligned}
mapmatching : \mathscr{L}(M) &\to \mathscr{L}(P), \\
(m_1, \ldots, mn) &\mapsto \underset{p_{iv_i} \in map(m_i)}{\arg\min} \quad eval(route(p_{1_{v1}}, p_{2_{v2}})) \mid \\
&\qquad \ldots \\
&\qquad route(P_{n-1,v_{n-1}}, P_{nv_n}))
\end{aligned}
\tag{1}
$$

Note, we require the output of *map* to be finite. This contradicts characteristics of real positioning systems as for each measurement an infinite number of possible positions may be the origin for this measurement. However, we can identify a *finite* set of representatives that are sufficient for our problem (see below).

Even though, we can easily denote the required function, its computation is difficult, if we have to consider execution time. A straight-forward approach would iterate through all permutations v_i to map a measurement m_i to a position p_{iv_i}. This, of course is virtually impossible for even small input sequences as the number of variations gets very large. Moreover, for every evaluation we have to call a time consuming *route* multiple times that executes optimal path planning on the road network.

2.2   Modelling Positioning System and Driver

The formalism above allows us to tailor the following approach to different scenarios that may differ in the used positioning system and characteristics of routes to detect.

Modelling the positioning system: Positioning systems are different according precision and distribution of measurement errors. A general model can be time-dependent; e.g. GPS errors depend on the satellite constellation. Usually a Gaussian distribution of measurements is assumed. For our approach the actual positioning model can be simplified as follows:

- The measurements to a position are distributed with a higher density closer to the actual position.
- There is a maximum error distance d_{max}.

The first point is obvious and is fulfilled by any Gaussian distributed positioning system. The second point is crucial: even precise positioning systems cannot state a maximum error – it is not intended to give a bound for any Gaussian distribution. However, as we have to keep the amount of possible mapped positions for a certain measurement finite, we have to state such a border. Even mathematically critical, this rule is not a problem in reality.

In our formalism, the positioning system model is reflected by two functions:

- *map*: only mapped positions up to the maximum distance d_{max} are computed;
- *eval*: the evaluation of routes takes into account the distances of measured and mapped positions.

Modelling the driver or driving goal: Drivers may have different opinions what a 'good' or 'typical' route is. Depending on the type of trip, the same driver can act different. E.g. the daily route to the office has a certain degree of optimality. In contrast the trip to watch some touristic sights or a travel to an unknown address may result in lower optimality measures and some sub-optimal bypasses. We can model this with the help of two functions:

- *route*: this function defines what generally is considered as 'good' or 'optimal'. One driver could consider the consumption of fuel as important, the other the driving time.
- *eval*: this function asserts a value that reflects the probability (in a wider meaning) that this route was actually driven. This function can reward certain attributes of the route (e.g. local optimality) to reflect the expected driving behaviour.

With these means we can model the system of roads and driver that affects the assumed route for a set of measurements.

2.3 Requirements, Assumptions

We want to declare some assumptions that we require for our approach later.

Assumption 1: We ignore the time of measurement. When collecting measurements, we could store the time stamp of the measured positions and we could consider these time stamps to evaluate of a route. The absolute values of time stamps are not of interest, but time differences of two subsequent measurements lead to the average speed in the driven section. We then could discuss about 'probable' speeds (e.g. in our *eval* function), however this is difficult. A speed can be arbitrarily low (e.g. red light or traffic jam). Thus, we can only identify unrealistic high speeds. This however would require to more carefully map a measurement to a position, in particular: an output of $map(m_i)$ depends on $map(m_{i-1})$ and $map(m_{i+1})$. This leads to an entirely different class of algorithms. As a starting point, in this paper we thus only discuss approaches that purely rely on the measured positions without a time stamp.

Assumption 2: We consider a finite number of variations per measurement. Figure 2 provides a motivation why the output of *map* is finite. If we look at Fig. 2a, a measurement can be mapped to an infinite number of positions that all reside on the same link. For better clearness, we only show the projection to one side of the road – our final implementation has to consider both driving directions.

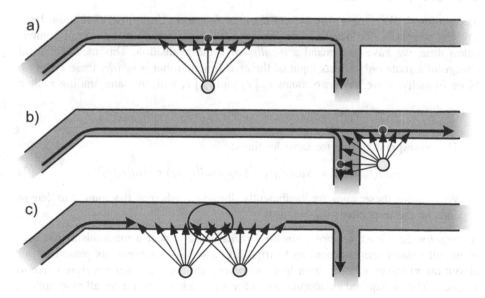

Fig. 2. Motivation to map measurements to finite sets of positions

In this example, the infinite set does not carry additional information: a route starting from left can be continued to the right with any of these positions, thus all positions can be represented by a single position, e.g., the closest to the measurement. More formally: all positions that reside on a certain link form an *equivalence relation*, and any representative is a useful mapping. In Fig. 2b we see that more than one equivalence class may result from a mapping, if possible positions reside on different links. However, we have to keep in mind:

- The considerations above do not apply for the very first and very last mapping of our measurements.
- If sets of possible mappings of subsequent measurements overlap, we cannot choose *arbitrary* representatives. If we chose positions inside the circle in Fig. 2c, resulting routes may include two U-turns, thus get significantly different.

We called the latter problem the *back-driving problem* and discuss it in more detail later. In simple words: if two measurements are too close, poorly selected representatives may result in routes of (partly) wrong direction.

Assumption 3: We use multi-routing as a building block. Even though our formalism only requires a *route* function for a single route, we strongly believe an efficient implementation requires a function

$$multiroute: \wp(P) \times \wp(P) \rightarrow \wp(R)$$

that computes *all* optimal routes between sets of start and target positions. Even tough such a function could call *route* for every permutation, we assume a more efficient approach as presented in [8].

Assumption 4: We expect a locality of the eval function. Until now, *eval* could have any characteristics. To benefit from some mechanisms to reduce the overall computation time, we have to demand a *locality* of route evaluation. This means: a local change of a route only affects input of the *eval* function that is *nearby* these changes. More formally: if we have two routes $r_a \mid r_x$ and $r_b \mid r_x$ with the same trailing r_x then

$$eval(r_a|r_x) < eval(r_b|r_x) \Rightarrow eval(r_a|r_x|r_y) < eval(r_b|r_x|r_y) \tag{2}$$

The corresponding rule for same leadings:

$$eval(r_w|r_x|r_a) < eval(r_w|r_x|r_b) \Rightarrow eval(r_x|r_a) < eval(r_x|r_b) \tag{3}$$

We require these rules for 'sufficiently long' r_x, whereas the minimum length depends on the respective *eval* function.

Assumption 5: We expect pre-segmented routes. We expect a mechanism that segments all measurements according to trips. I.e. a set of measurements passed to our algorithm represent, e.g. trip *from home to office*, another set represents *from office to shop* etc. This is required as stopping a trip for e.g., parking invalidates all assumptions about routes, e.g. regarding optimality. In particular, any new route goes to any direction regardless from prior driving. The basis of a segmentation mechanism is to detect halts of steady positions; however, such a mechanism in not part of this paper.

Strongly related to this issue are U-turns. We define U-turns as a $180°$ turnaround on the *same* road. If we expect finite mappings per measurements (assumption 2), we have a certain exception for start and end of routes. As U-turns can be viewed as end of route and start of a new route, we would thus produce this exception at every occurrence. Even though we could access minus points for U-turns in the *eval* function, we decided to directly filter out variations with U-turns to reduce the overall number of variations. To deal with U-turns in real routes, we have to rely on a segmentation mechanism to cut the sequence of measurements. Note that more complex manoeuvres to turn the direction (e.g. 'three left turns around the block') are not crucial.

2.4 Problems, Failures

The offline map matching problem turned out to be surprisingly hard to solve. The actual research was executed over two years whereas several ideas were realized that turned out not to be successful. We identified two major problems.

Problem 1: Combinatorial explosion of variations. If we mapped a single measurement to multiple possible positions and connect them, the resulting set of variations get huge very quickly. An established approach to deal with such a problem is *Viterbi* [10]. It tries to detect hidden states (real positions) from observed events (our measurements). It implicitly deals with a huge number of variations as it only considers the most probable last state for a new hidden state, thus the number of variations does not increase. However, this is a significant problem, as a decision will not be revised later, when we know about the subsequent states. Thus, the results were usually poor.

We also tried *Simulated Annealing* [1] – an approach to find a global optimum of a function (here *eval*). This also was a failure. It expects for small changes in the input also small changes in the output. However, changing a mapping of a measurement can produce arbitrary big changes in the evaluation.

Another experiment was a *Divide-and-Conquer* approach: we first check optimal routes from all start to target measurement mappings. If one of these routes approximate all intermediate measurements, we have a solution. If not, we look for an intermediate measurement to split and recursively try to find an appropriate route. This approach again was a failure, whenever the route was not driven according to optimality measures assumed by our *route* function. As a result, the recursive segmentation goes very deep and produces a combinatorial explosion we tried to avoid. Moreover, the approach to independently search for appropriate routes failed as the route parts are not actually independent – at the intermediate points, the values depend on both parts of connected routes.

As a final direction to solve this problem we tried to directly check certain properties of possible routes. If not fulfilled, we exclude them without asking the *eval* function. This significantly reduced the set of possible route variations. We conducted experiments with the property of *local target orientation* [6]. This property means: for a sufficiently small part of every trip we can find an optimal route, even if the overall route is not optimal. This approach also was a failure due to two reasons: first, the computation costs to check the local target orientation were considerable high. Second and most important: the degree of locality depends on the current driving situation.

If we used a fix value, very often the set of possible routes gets empty, as not any route fulfils the given property.

Problem 2: The back-driving problem. This occurs as measurements can be very close and suggest a 'driving in the opposite direction' (Fig. 3). In our example the route goes from left to right (Fig. 3a). The route positions Pos_{10} and Pos_{11} were mapped to measurements M_{10} and M_{11}. Even though they are in range of the maximum error d_{max}, they change their ordering according to the route direction. If we now map the measurement to the nearest positions on the road network (p_{10j}, p_{11k}), a possible route extension has to drive a '*route around the block*' (Fig. 3b; the numbers ①, ② and ③ indicate the driving sequence of the assumed route). Obviously the assumed route significantly differs from the real route, even though the mapped positions are very close to the real positions.

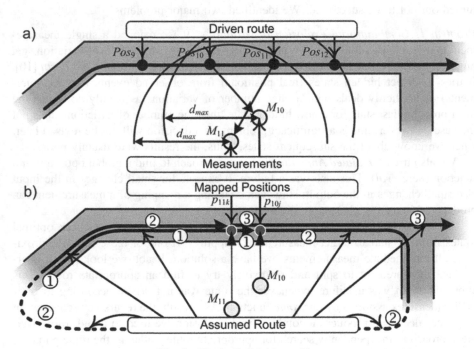

Fig. 3. The back-driving problem

2.5   An Efficient Solution

Considering the problems above, we finally developed a solution for the offline map matching problem. It iterates through the measurements from trip start to termination. For each iteration we extend all conceivable route variations by new variations that result from possible mappings of a measurement.

Figure 4a shows the situation after three steps. We manage a set V that holds all conceivable route variations through the mapped positions. If we take the next

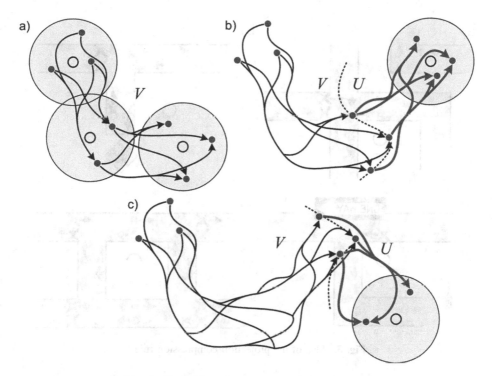

Fig. 4. Basic idea of our map matching approach

measurement (Figs. 4b and c), we create all optimal routes (U) from current route end-points to mapped positions of the new measurement. For the next iteration, we now have to create a new set V, built from all extensions of old V extended with all routes of U.

Due to the high number of route variations, we have to incorporate mechanisms to reduce computing costs to a manageable level.

1. Mechanisms to reduce the number of mappings. According to assumption 2, we map a measurement to the nearest position of all links in range. The easiest way to reduce the number of route variations that result from different mappings is to reduce the mappings themselves. Our mechanism (1a) is not to map all measurements but only a reasonable subset. The motivation: the number of measurements is pre-defined by the position measurement system, e.g. by the smart phone app that logs the trip positions. This could, e.g., measure one position per second. This set up is not under control by the map matching system, but *given*. Our algorithm should thus have the chance to take into account a lower number of measurements to produce route variations. As we thus can keep a certain minimum distance between measurements, this is an effective mechanism against the back-driving problem. In our approach, we do not ignore the other measurements but always evaluate *all* measurements in the *eval* function.

Our second mechanism (1b) is called the *projection compression rule*. Figure 5a shows a scenario where a measurement produces 10 mapped positions. There exist 14

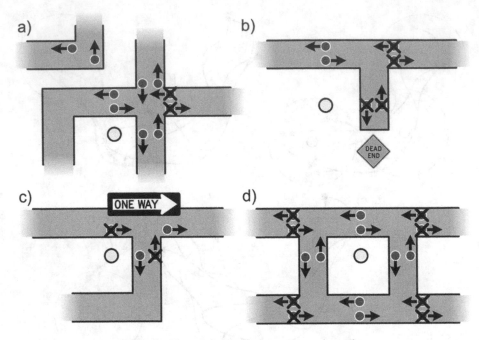

Fig. 5. Idea of the projection compression rule

routes to drive through the range of this measurement (2 through the upper left corner, 12 = 4·3 through the crossing). However, we could remove two of the 10 mappings (marked with 'X') without loosing any of the 14 route variations.

Also dead ends and one way roads may reduce the mappings without loosing variations (Fig. 5b and c). Figure 5d shows a complex situation with 24 variations to drive through (without counting loops). Here, we can remove 8 of the 16 mappings without loosing variations.

To identify removable mappings we proceed as follows:

- We remove all mappings in dead ends.
- For each remaining mapping, we assign the distance of measurement and mapping. If for a mapping p_i all links to drive *to* or drive *from* p_i contain a nearer mapping p_j, we can remove mapping p_i.

Note that the projection compression rule does not avoid all double variations. E.g. in Fig. 5a, the routes through the 4-way-crossing from top to bottom are still presented by two measurements.

As a last mechanism (1c) to reduce the number of mappings: we can remove all mappings that are not endpoint of any currently considered route variation.

2. Mechanisms to reduce the number of route variations. This first mechanism (2a) is to directly avoid U-turns and not to wait for a bad evaluation by the *eval* function later (see assumption 5). Our approach iterates through measurements and tries to extend existing route variations by a new part of the route. Thus, the only chance to produce a

U-turn is when an existing route is extended. As a simple mechanism, we only accept extensions without U-turns.

A second mechanism (2b) 'compresses' the set of existing variations. The idea: if an existing route variation is not able to be part of the final route with the best evaluation, we can remove it. We apply assumption 4, formula (2) for this. In more detail: if we identify two route variations with the same trailing r_x, the one with the worse evaluation cannot be the final 'winner' ($r_b \mid r_x \mid r_y$ according to formula (2)). We thus only have to check all variations for same trailings, evaluate them and keep only the best.

3. Mechanism to speed up the evaluation. For each iteration and variation, *eval* has to be executed again. This is because old results of eval are not useful anymore as routes are extended (by a route of *U*). Even though the number of variations stays constant (due to mechanisms 2a and b), the length of each route gets longer as more measurements have to be taken into account. As a result, for a naïve realization, the execution time of *eval* would significantly increase for each iteration.

As our final mechanism we apply assumption 4 and formula (3). It turned out that after a certain time, all route variations in *V* have a same leading route ($r_w \mid r_x$ in formula (3)). Whenever this happens, we can ignore r_w for new evaluations as it does not affect the ordering.

We now can put our considerations together to our map matching algorithm below (references to the mechanisms are in brackets). It contains two main loops: one to compute all mappings and one to subsequently extend route variations.

Note that in contrast to the formal definition (1), this algorithm returns a route, not a list of mapped positions. However positions and a route that connects these positions are equivalent information.

mapmatching: input: measurements m_i	
output: most probable route v	
select a subset $\mu=\{\mu_1,\ldots,\mu_m\}\subseteq\{1,\ldots,n\}$, with $\{1,n\}\subseteq\mu$	(1a)
for each $\mu_i\in\mu$ {	
compute $P_i= map(m_{\mu_i})$	
compress P_i according *projection compression rule*	(1b)
}	
$V=multiroute(P_1, P_2)$	
for $i=3$ to m {	
$P'_{i-1}=\{p\in P_{i-1} \mid p$ *is endpoint of a* $v\in V\}$	(1c)
$U=multiroute(P'_{i-1}, P_i)$	
$V= \bigcup_{v\in V, u\in U} v \mid u$ *if* u *extends* v *without producing a U-turn*	(2a)
compress V	(2b)
}	
return $\arg\min_{v\in V} eval(v)$	(3)

2.6   Again the Back-Driving Problem

According to Figs. 2c and 3 the back-driving problem can occur, if two subsequent measurements have overlapping circles (with radii d_{max}) and at least one link resides in the overlapping area. With the selection of the subset μ we are able to control this: we can either select measurements that have at least a distance of $2d_{max}$ (no overlapping areas at all), or we can geometrically check for links in the overlapping areas. The latter approach requests some complex geometric computations, but tolerates closer measurements.

However, for positioning systems with large positioning errors, this selection causes a low measurement density. The problem: if two selected subsequent measurements are too far, the route in-between is not represented adequately.

To solve this, we introduced another approach that allows to select arbitrary close measurements with the costs of additional computation. We modify the *map* function: a position is not mapped to the nearest position on a link but on a fix *relative* position, e.g. always on the centre of the link's running length. As a result: even close measurements do not produce back-driving, as subsequent mapped positions stay constant on a link. This is possible due to assumptions 1 and 2. Two crucial points:

- The very first and very last measurement should additionally be mapped to nearest route points.
- We have to deactivate the projection compression rule, otherwise we still may produce the back-driving problem. This may occur, if nearby a crossing, the rule removes projections in the wrong driving order.

If we consider more measurements, we have a longer runtime due to more multi-route calls. In addition, the number of mappings per measurement is increased on average, thus this modification may dramatically increase the overall runtime.

A last consideration: selecting two measurements that are closer than $2d_{max}$ would allow us to detect small route changes between measurements that were otherwise undiscovered. On the other hand, these route changes may be below the measurement precision, thus remain undiscovered even though respective measurements are selected. As a best practice, we thus suggest at least $2d_{max}$ distance between selected measurements.

2.7   A Best-Practice Eval Function

We currently only request assumption 4 to be fulfilled by the *eval* function. In this section we describe an *eval* function that turned out to be useful for a driving scenario.

Our function considers the following properties (Fig. 6):

- The total route length L in meters.
- The sum of squares of distances d_i between measurements and nearest route positions scaled by $1/d_{max}^2$.
- The square of first route point to first measurement d_s and last route point to last measurement d_t scaled by $1/d_{max}^2$.

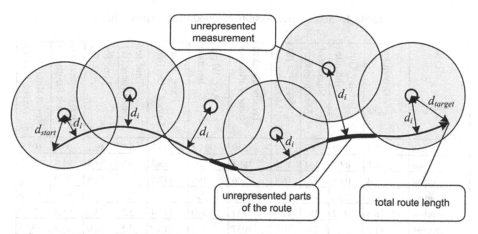

Fig. 6. Properties considered by the evaluation

- The route meters not represented by any measurement L_{unrep}, and the number of measurements not covering a route position cnt_{unrep}.

We build the weighted sum of these values (4).

$$eval(r) = \begin{pmatrix} L \\ \sum d_i^2/d_{max}^2 \\ (d_s^2 + d_t^2)/d_{max}^2 \\ L_{unrep} \\ cnt_{unrep} \end{pmatrix} \cdot \begin{pmatrix} w_{len} \\ w_{dist} \\ w_{st} \\ w_{lunrep} \\ w_{cntunrep} \end{pmatrix} \tag{4}$$

We achieved good results with the settings: $w_{len} = 1/3$, $w_{dist} = 50$, $w_{st} = 100$, $w_{lunrep} = 1$, $w_{cntunrep} = 10$.

3 Evaluation

We fully implemented the presented offline map matching approach inside our routing platform *donavio* [6] that is part of the *HomeRun* environment [5, 7]. We conducted a number of experiments to show the efficiency of our computations and the quality of the output. For all routes we set $d_{max} = 200$ m.

We selected measurements with a distance of $2d_{max}$. For execution time measurements we used a PC with i7-4790 CPU, 3.6 GHz. Table 1 summarizes the results; some routes are illustrated in Fig. 7.

To assess the similarity of the assumed route by our algorithm and driven route, we measured the amount of original positions that reside on the result route. As the result route may also contain additional tracks, we also compared the route lengths. In summary, in all our experiments, the result routes were always very similar to the original driven route.

Table 1. Summarized Results of Conducted Experiments

Type	Driven km	Driving Time (hh:mm)	Number of Measurements	Mappings per Measurement	Totally Checked Route Variations	Avg. $|V|$ per Iteration	Exec. Time/ Mapped Mes. (ms)	Time for *multiroute* (%)	Time for *eval* (%)	Org. Positions on Result Route (%)	Result Route km/ Driven km (%)
urban	2.3	0:07	69	13.7	510	23	108.7	8.7	40.1	100.0	105.9
urban	4.8	0:07	151	17.5	1891	90	315.2	4.8	81.8	95.4	97.7
urban	5.4	0:08	129	16.3	2306	105	445.1	3.4	86.5	96.9	99.8
urban	16.1	0:32	458	6.9	2229	294	78.7	2.2	54.1	97.2	100.5
urban	18.5	0:22	186	12.8	4335	290	177.4	5.5	73.7	98.4	99.7
suburban	11.2	0:14	240	10.1	2037	139	113.9	3.5	66.0	95.4	102.3
suburban	16.1	0:21	301	10.5	3108	238	272.9	2.3	84.6	99.7	99.1
suburban	19.2	0:22	457	6.9	1776	188	94.6	2.9	61.3	97.4	99.3
interurban	412.3	3:39	4124	9.5	38377	3492	454.7	2.0	76.8	98.0	100.0
interurban	977.8	8:42	9650	9.8	112458	8425	550.0	1.2	59.5	99.0	98.7

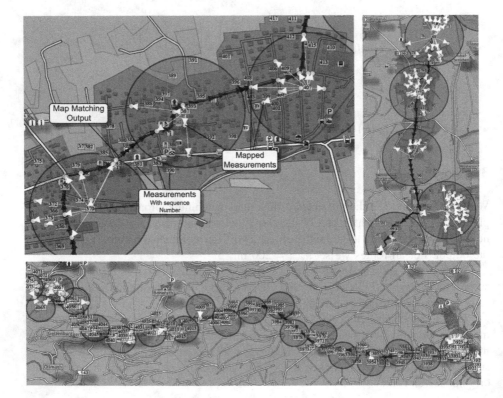

Fig. 7. Example measurements and result routes

We also measured the execution time. Even though it is not primarily required for an offline approach to perform execution in time, we got approx. 260 ms per mapped measurement. Thus, this approach could in principle also process measurements in realtime.

We tried to identify the component that needed most of the execution time. We thus analysed the CPU usage for the *multiroute* and *eval* functions. The multiroute function only requires 4 % on average. This supports assumption 3 to heavily rely on the multi-routing function. We also conducted experiments where we replaced our multi-routing approach by permutations with single routing (not shown in the table). The result: our multi-routing is factor 1022 times faster (max. 2355 times) compared to a single-routing approach.

The most time-consuming function in our implementation is the *eval* function. This is due to complex geometric computations, in particular to identify the unrepresented parts of the route. For time-critical execution scenarios we could consider to simplify the *eval* function.

As a last set of experiments, we also checked the alternative mapping for measurements closer than $2d_{max}$ (Sect. 2.6). We thus also select measurements with a distance of d_{max}. However, disabling the projection compression rule increases the number of mappings per measurement by factor 2.5 on average; as expected this significantly affects the overall computation time: it increased by factor 4.2.

4 Conclusions

In this paper we presented an approach for offline map matching. The main idea was to map a measurement to all potential road positions and to check, how a sequence of mapped positions can be connected by a driven route. It heavily relies on a multi-routing mechanism to immediate check connections between set of start and target positions. To address the problem of combinatorial explosion, we suggest a number of mechanisms. An evaluation shows the effectiveness of these mechanisms.

References

1. Bertsimas, D., Tsitsiklis, J.: Simulated Annealing. Stat. Sci. **8**(1), 10–15 (1993)
2. Haunert, J.H., Budig, B.: An algorithm for map matching given incomplete road data. In: Proceedings of the 20th International Conference on Advances in Geographic Information Systems, pp. 510–513 (2012)
3. Marchal, F., Hackney, J., Axhausen, K.W.: Efficient map-matching of large GPS data sets – tests on a speed monitoring experiment in Zurich. J. Transp. Res. Board **2005**, 93–100 (1935). Transportation Research Re-cord
4. Pereira, F.C., Costa, H., Pereira, N.M.: An off-line map-matching algorithm for incomplete map databases. Eur. Transp. Res. Rev. **1**(3), 107–124 (2009)
5. Roth, J.: Combining symbolic and spatial exploratory search – the homerun explorer. In: Innovative Internet Computing Systems (I2CS), Hagen, June 19–21 2013, Fortschritt-Berichte VDI, Reihe 10, Nr. 826, pp. 94–108 (2013)

6. Roth, J.: Predicting route targets based on optimality considerations. In: International Conference on Innovations for Community Services (I4CS), Reims (France) 4–6 June, 2014, pp. 61–68. IEEE xplore (2014)
7. Roth, J.: Generating meaningful location descriptions. In: International Conference on Innovations for Community Services (I4CS), July 8–10, 2015, Nuremberg (Germany), pp. 30–37. IEEE xplore (2015)
8. Roth, J.: Efficient many-to-many path planning and the traveling salesman problem on road networks. In: KES Journal: Innovation in Knowledge-Based and Intelligent Engineering Systems, to appear (2016)
9. Schuessler, N., Axhausen, K. W.: Map-matching of GPS traces on high-resolution navigation networks using the Multiple Hypothesis Technique (MHT), Working Paper 568, Institute for Transport Planning and System (IVT), ETH Zurich (2009)
10. Viterbi, A.: Error bounds for convolutional codes and an asymptotically optimum decoding algorithm. IEEE Trans. Inf. Theory **13**(2), 260–269 (1967)
11. White, C.E., Bernstein, D., Kornhauser, A.: Some map matching algorithms for personal navigation assistants. Transp. Res. Part C Emerg. Technol. **8**, 91–108 (2000)
12. Yanagisawa, H.: An offline map matching via integer programming. In: 2010 Proceedings of the 20th International Conference on Pattern Recognition (ICPR), pp. 4206–4209. IEEE (2010)
13. Zhou, J., Golledge, R.: A three-step general map matching method in the GIS environment: travel/transportation study perspective. Int. J. Geogr. Inf. Syst. **8**(3), 243–260 (2006)

Using Data as Observers: A New Paradigm for Prototypes Selection

Michel Herbin$^{(\boxtimes)}$, Didier Gillard, and Laurent Hussenet

CReSTIC, Université de Reims Champagne-Ardenne, Institut of Technology,
Chaussée du Port, 51000 Chalons-en-champagne, France
{Michel.Herbin,Didier.Gillard,Laurent.Hussenet}@univ-reims.fr
http://crestic.univ-reims.fr

Abstract. The prototype selection is a bottleneck for lot of data analysis procedures. This paper proposes a new deterministic selection of prototypes based on a pairwise comparison between data. Data is ranked relative to each data. We use the paradigm of the observer situated on the data. The ranks relative to this data gives the viewpoint of the observer to the dataset. Two observers provide a link between them if they have no data between them from their respective viewpoints. The links are directed to obtain a directed graph where data is the set of vertices of the graph. The observers move using the directed graph. They reach a prototype when they arrive at a viewpoint with no outgoing connexion of the directed graph. This method proposes both the prototype selection and the structuration of the dataset through the directed graph. The paper also presents an assessment with three kinds of datasets. The method seems particularly useful when the classes are hardly distinguishable with classical clustering methods.

Keywords: Prototype selection · Instance selection · Case-based reasoning · Dimension reduction

1 Introduction

The prototype selection (i.e. the instance selection) consists in selecting a subset of data from a dataset in a way that the subset could represent the whole dataset. It is a bottleneck for lot of data analysis procedures. The prototype selection is a crucial step in big data classification [1], in machine learning [2], in data mining [5], in nearest neighbour supervised classification [3,4], in medical case-based reasoning [6], in social recommender system [7],... Also one can find a review of different methods of instance selection in [8,9].

In the context of dataset exploration, each data is a case that we examine to better understand the data sample. This approach is tedious and the user must reduce the number of data he wishes to examine. For this purpose he selects some prototypes. Most of methods of prototype selection are associated with clusters. Except the random search, we do not know a method of selection of prototypes without reference to a clustering of the dataset. Unfortunately

© Springer International Publishing AG 2016
G. Fahrnberger et al. (Eds.): I4CS 2016, CCIS 648, pp. 39–46, 2016.
DOI: 10.1007/978-3-319-49466-1_3

the clusters are unknown at the first step of data exploration and the classical methods of clustering make implicit assumptions on the number or the statistical properties of the eventual clusters. In this paper we propose a new deterministic selection of prototypes without any clustering approach. The method is only based on a pairwise comparison between data.

First we expose the paradigm of observer using a pairwise comparator. Then we describe the design of the network linking data. The observers move to the prototypes using this network. Second we assess the method of prototype selection with three kinds of datasets. Finally we conclude this paper by indicating a specific usefulness of our method.

2   Dataset and Observers

2.1   Pairwise Comparisons and Rankings

Let Ω be a set of n data:

$$\Omega = \{X_1, X_2, ...X_n\}$$

This set is provided with a similarity index which allows pairwise comparisons between data.

The pairwise comparisons have been studied for a long time in the literature [11,12]. Many similarity indices are directly based on these pairwise comparisons. The similarity indices are often derived from distances or a pseudo-distances, and many methods are developed to build such similarity indices based on pairwise comparisons [10]. In this paper, we use only some basic properties of a similarity index.

Let S be the similarity index. Let A be a data in Ω. The index should allow all the comparisons relative to A. Thus we need to have the following properties:

- $S(A, A) \geq S(A, B)$,
- $S(A, B) \geq S(A, C)$ if B is more similar to A than C is,

where $A \in \Omega$, $B \in \Omega$, and $C \in \Omega$.

These properties allow to define a total pre-order relation in Ω relative to the data A. This relation is called $\leq_A$ and it is defined by:

$$B \leq_A C \quad \text{iff} \quad S(A, B) \geq S(A, C)$$

The dataset induces one ranking per data. Then we have n rankings. The ranking induced by X_i in Ω is defined by:

$$rk_i = (rk_{i,1}, rk_{i,2}, rk_{i,3}, ...rk_{i,n})$$

where $rk_{i,j}$ is the rank of X_j with the relation $\leq_{X_i}$. Note that rk_i is a permutation on $1, 2, 3, ...n$ and $rk_{i,i} = 1$.

In this paper, we use the paradigm of the observer. Each observer is situated on a data which gives a viewpoint of Ω. Thus we consider that the ranking rk_i

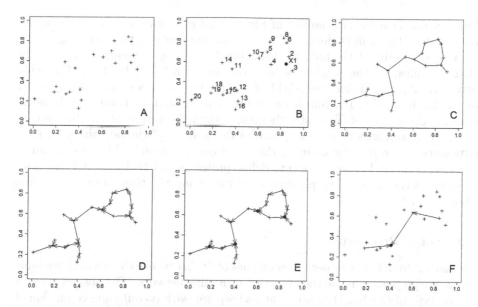

Fig. 1. Processing pipeline for selecting prototypes: A. Simulated dataset in dimension 2 ($n = 20$), B. Ranking relative to X1 using Euclidean distance as pairwise comparator, C. Network with links between data, D. Directed graph from network, E. Selection of four prototypes (black circles), F. Second run for selecting one prototype from four previous ones.

is the viewpoint of Ω from X_i and data is ranked from the first to the last at each viewpoint. Figure 1A displays an example of a two dimensional simulated dataset with $n = 20$. Figure 1B shows the ranking relative to a data X_1 using Euclidean distance. This ranking gives the viewpoint on the dataset from X_1.

2.2 Links and Network

In this section we select a subset of representatives of Ω using this paradigm of observers. First we define links between data. Then we use these links to move the observers to the optimal viewpoints.

A link between the viewpoints X_i and X_j is obtained when we have no step to move between X_i and X_j. Let us define a step between X_i and X_j. Let us consider the viewpoint from X_i and the viewpoint from X_j. The data X_k is a step between X_i and X_j when it lies both between X_i and X_j using the viewpoint X_i and between X_j and X_i using the viewpoint X_j. In other words X_k sets a possible step to go between X_i and X_j when we have:

$$rk_{i,i} < rk_{i,k} < rk_{i,j} \text{ and } rk_{j,j} < rk_{j,k} < rk_{j,i}$$

If no step is possible between X_i and X_j, then we define a link between X_i and X_j. Thus the observers X_i and X_j provide a link if they see no common data from their respective viewpoints. Figure 1C displays an example of the links we obtain using the dataset of Fig. 1A and Euclidean distance as pairwise comparator. These links define a network that we study in the following.

In order to direct the links of this network, we assign to each data a weight that is defined by the number of links coming from the data itself. The more attractive X_i, the higher the weight of X_i (i.e. the number of links going to X_i or coming from X_i). The link between two data X_i and X_j is directed from X_i to X_j if the weight of X_j is strictly greater than the weight of X_i. In the case of equality, we consider the maximum of the weights of the neighbors for each extremity of the link for defining the direction of this link. Thus we obtain a directed connected graph we use to define prototypes of Ω. Figure 1D shows an example of the directed graph that we obtain using the procedure we describe above.

2.3   Set of Prototypes

Our goal is to define a subset of prototypes of Ω. Let us use the network we built on Ω. We propose to move the observers using this network. The observers reach a prototype of Ω when they arrive at a viewpoint with no outgoing connexion of the network. The subset of prototypes is called Ψ. It gives the prototypes that we propose for reducing the dimension of Ω. If the number of prototypes is too large, then the procedure reducing the dimensionality of Ω can be iterated on the subset of prototypes. Figure 1E displays the results of a first run of our prototypes selection on a simulated dataset of Fig. 1A, we obtain four prototypes (black circles). Figure 1F displays the results of a second run based on the previous results. This second run gives only one prototype.

3   Experimental Study

This section is devoted to the study of our method of prototype selection. The prototypes are typically associated with the classes that form a partition of the dataset. Ideally the algorithm of prototype selection should give one and only one prototype per class. Unfortunately the concept of classes has no meaning in the framework of exploratory analysis when the classes and the number of classes are unknown. Moreover the classes have no use for many applications. For instance, the use of a priori clusters for the classification of users or information could be against productive to give dynamic personalized recommendations in the framework of recommender systems. In this context, the prototype selection is connected with the structuration of the dataset using a network. The main advantage of our method is to propose both the prototype selection and the structuration of the dataset through a directed graph.

In this paper, we place ourselves resolutely in the context of the exploratory analysis of data without any a priori assumption on eventual classes, we only use an index of pairwise comparison. But the use of classes gives the most classical way to evaluate a set of prototypes. So this paper uses classes to study and to

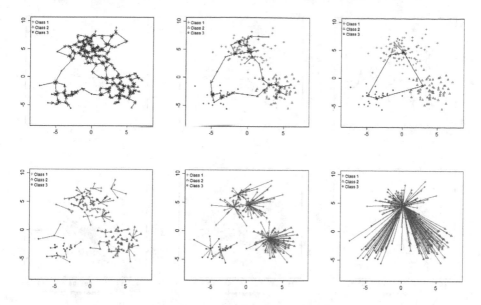

Fig. 2. Selection of prototypes using a simulated dataset ($n = 200$) with three normal distributions (100, 70, 30). Top: Iterative selections of prototypes with respectively 34, 6, and 1 prototypes (black circles). Bottom: Respective links between data and their associated prototypes.

assess only the prototype selection, the detection of classes (i.e. the clustering) is out of the scope of this paper.

First we present the assessment of our method of prototype selection using distinguished classes for ease of understanding. Second we display the result we obtain using the classical IRIS dataset. Then we propose an example when the classes are hardly distinguishable with classical clustering methods.

3.1   Assessment of a Set of Prototypes

In the following we simulate a dataset with three classes to which we apply our method of selection of prototypes. The set of prototypes is assessed using the labels of the classes. Each data is assigned to the prototype which is the most similar to itself with respect to the pairwise comparator. Thus each data X_i in Ω is assigned to the prototype P_j in Ψ where P_j has the smallest rank of all the prototypes of Ψ relative to X_i. P_j is the closest prototype from the viewpoint X_i. Then we define δ_i which is equal to 1 if X_i and P_j have the same label (i.e. they belong to the same class) and 0 otherwise. If δ_i is equal to 1, then we say that X_i is well represented by the set of prototypes. Otherwise X_i is not represented by the set of prototypes. The percentage of data in Ω that is well represented by Ψ defines our criterion to assess Ψ. When the percentage is equal to 100 %, the set of prototypes Ψ can represent the whole dataset.

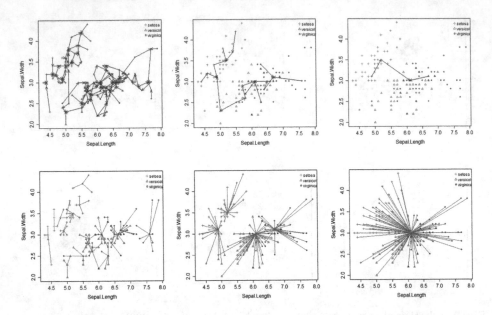

Fig. 3. Display of Iris dataset ($n = 150$, three classes of 50 data, in dimension 4) projected on two components: sepal length and the sepal width. Three iterations of prototype selection using Euclidean distance in dimension 4: 27, 6 et 1 prototypes are selected, 94.7 %, 88 %, 33.3 % of data are respectively well represented by these prototypes.

For understanding the criterion, we select Ψ using a dataset with three well separated clusters. The clusters are simulated using Gaussian distributions in dimension 2. Ω has 200 data. The clusters have 100, 70 and 30 data. Figure 2 displays three runs of our prototypes selection using Euclidean distance as pairwise comparator. We obtain 34, 6 and 1 prototypes (Fig. 2, first row) Each data is associated with the most similar prototypes (Fig. 2, second row). The percentages of well represented data by 36, 6, and 1 prototypes of the three iterative runs are respectively 98.5 %, 99 %, and 50 %. We note that the percentage of well represented data does not directly depend on the number of prototypes.

3.2   Prototype Selection with a Real Dataset

In the following we use the classical *IRIS* dataset of Machine Learning Repository of UCI [13]. The set has 150 data with three classes of 50 data each: Iris Setosa, Iris Versicolour, Iris Virginica. Each data is given in dimension four with sepal length in cm, sepal width in cm, petal length in cm, and petal width in cm.

Figure 3 displays the iris data using only two components: the sepal length and the sepal width. Three iterations of our prototype selection are successively applied in this dataset in dimension four. We obtain respectively 27, 6 and 1 prototypes that respectively represent 94.7 %, 88 %, and 33.3 % of data.

3.3   Hardly Distinguishable Classes and Prototypes

In the following we simulated a dataset with two classes that are hardly distinguishable when using classical clutering methods. Indeed classical clustering methods are often based on statistics such as means or medoids. They use these statistics to determine the prototypes and they make the assumption that data could be well represented by the selected prototypes. Unfortunately these statistical approaches are often unadapted.

Figure 4 presents a datset ($n = 300$) where two classes that are simulated using an uniform distribution. The 300 data are simulated in dimension 2 within two rectangular crowns. The two classes are hardly distinguishable when using classical clustering methods without making any assumption about the shapes of the clusters. The classical clustering methods (k-means, k-medoids) fail even after a transform with Principal Component Analysis (PCA). Figure 4 displays the results of three iterative selections of prototypes. We select first 58 prototypes with 100 % of well represented data, second 12 prototypes with 87.6 % of well represented data, and third 2 prototypes with 29.3 % of well represented data.

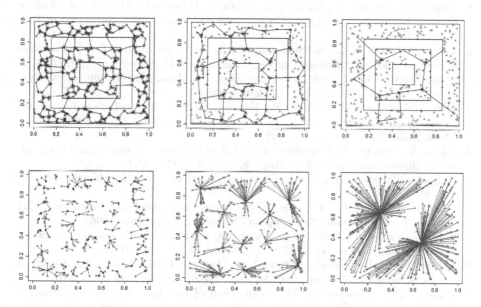

Fig. 4. Dataset ($n = 300$) with a uniform distribution in two rectangular crowns and three iterative selection of prototypes with respectively 58, 12, and 2 prototypes and 100 %, 87.6 % and 29.3 % of well represented data.

4   Discussion and Conclusion

The method we propose for selecting prototypes is deterministic. It does not require to establish a number of prototypes. It does not assume any assumption on the shape and the number of eventual clusters. Moreover it gives a structuration of the dataset through a directed graph. So we propose the use of this method in a first approach of exploratory analysis when we have only a pairwise comparator and no knowledge of the number and possible properties of eventual clusters. This kind of dataset exploration through prototypes selection becomes essential when the clusters are indistinguishable, such as when the clusters have a strong overlap.

The method we propose is not optimized in the version presented in this paper. It only gives a first approach to reduce the size of a dataset. Its use for large datasets is to be developed in future work.

In this paper we use the assignment of a data to a prototype only for assessment of our method. But the procedure for assigning a data to a prototype should also be improved for studying how to update the network when a new data is added to the dataset.

References

1. Triguero, I., Peralta, D., Bacardit, J., Garca, S., Herrera, F.: MRPR: a mapreduce solution for prototype reduction in big data classification. Neurocomputing **150**, 331–345 (2015)
2. Lesot, M.-J., Rifqi, M., Bouchon-Meunier, B.: Fuzzy prototypes: from a cognitive view to a machine learning principle. In: Bustince, H., Herrera, F., Montero, J. (eds.) Fuzzy Sets and Their Extensions: Representation, Aggregation and Models, vol. 220, pp. 431–452. Springer, Heidelberg (2007)
3. Garcia, S., Derrac, J., Cano, J.R., Herrera, F.: Prototype selection for nearest neighbor classification: taxonomy and empirical study. IEEE Trans. Pattern Anal. Mach. Intell. **34**(3), 417–435 (2012)
4. Nanni, L., Lumini, A.: Prototype reduction techniques: a comparison among different approaches. Expert Syst. Appl. **38**, 11820–11828 (2011)
5. Reinartz, T.: A unifying view on instance selection. Data Min. Knowl. Disc. **6**, 191–210 (2002)
6. Schmidt, R., Gierl, L.: The roles of prototypes in medical case-based reasoning systems. In: 4th German Workshop on Case-Based Reasoning (1996)
7. Song, Y., Zhang, L., Giles, C.L.: Automatic tag recommendation algorithms for social recommender systems. ACM Trans. Web **5**(1), 1–34 (2011)
8. Olvera-Lopez, A., Carrasco-Ochoa, J.A., Martinez-Trinidad, J.F., Kittler, J.: A review of instance selection methods. Artif. Intell. Rev. **34**, 133–143 (2010)
9. Jankowski, N., Grochowski, M.: Comparison of instances seletion algorithms I. Algorithms survey. In: Rutkowski, L., Siekmann, J.H., Tadeusiewicz, R., Zadeh, L.A. (eds.) ICAISC 2004. LNCS (LNAI), vol. 3070, pp. 598–603. Springer, Heidelberg (2004). doi:10.1007/978-3-540-24844-6_90
10. Cunningham, P.: A taxonomy of similarity mechanisms for case-based reasoning. IEEE Trans. Knowl. Data Eng. **21**(11), 1532–1543 (2009)
11. Thurstone, L.: A law of comparative judgment. Psychol. Rev. **34**, 273–286 (1927)
12. Bellet, A., Habrard, A., Sebban, M.: A survey on metric learning for feature vectors and structured data. Technical report (2014). arXiv:1306.6709
13. Bache, K., Lichman, M.: UCI Machine learning repository. University of California, Irvine, School of Information and Computer Sciences (2013). http://archive.ics.uci.edu/ml

Monitoring and Decision Making

Reconstruct Underground Infrastructure Networks Based on Uncertain Information

Marco de Koning[1,2] and Frank Phillipson[1(✉)]

[1] TNO, The Hague, The Netherlands
frank.phillipson@tno.nl
[2] VU University, Amsterdam, The Netherlands

Abstract. This paper focuses on developing methods to reconstruct the physical path of underground infrastructure networks. These reconstructions are based on the structure of the network, start and endpoints and the structure of an underlying network. This data is partly considered unreliable or uncertain. Two methods are presented to realise the reconstruction. The first method finds a path of given length in a directed graph by applying a modified version of Yen's algorithm for finding K-shortest simple paths in a directed graph. A second, so-called Bottom-up approach is developed which aims to take advantage of the structure of the underlying network. The developed methods are applied on a series of examples for comparison.

Keywords: Underground infrastructure · K-shortest path problem · Reconstruction

1 Introduction

Most underground networks to deliver services such as electricity, gas, telecommunications and television were built during the 20th century. The administrations of these networks were done by hand resulting in a huge collection of (paper) maps. In the last two decades of that century, network administrators started to digitise the maps, first by scanning them, later by drawing the networks in CAD-like software. Unfortunately, a huge part, especially the dense last parts of the networks, close to the houses, is still not digitised or only scanned. However, in some of these cases, some of the nodes of the network, the structure of the network and distances are known. In this paper is tried to reconstruct the physical path of the network based on this information, where is included that some of this information is unreliable or uncertain. The reason for this unreliability and uncertainty are mistakes, caused by all the work by hand in earlier processes, or estimations, e.g., the length of parts of the network that can be based on measurements. As far as the authors know, this problem was not described before.

In this paper the methodology for the reconstruction is demonstrated on an actual network, namely the copper network used for telecommunication services.

© Springer International Publishing AG 2016
G. Fahrnberger et al. (Eds.): I4CS 2016, CCIS 648, pp. 49–58, 2016.
DOI: 10.1007/978-3-319-49466-1_4

In order to supply end users with higher bandwidth, the network is developed into an Fibre to the Curb (FttCurb) network, where the, presumed, existing fibre optic network of Fibre to the Cabinet (FttCab) is extended further into the remaining copper network. The new active nodes for the FttCurb network are logically placed on splicing points of the copper network (see, e.g., [12]). Here is assumed that FttCab has been completely implemented and the focus lies on the FttCurb roll-out. A problem, however, is that the precise location of the splices is not known. Information is available on scanned maps (ergo, not available for extensive automatic use) and the available digital information is not 100% reliable.

The structure of the paper is as follows. First, the problem description is explained in more mathematical detail and some preliminary notation is introduced. Then, some pre-processing is presented in order to prepare the data to be used. Next the complete description and analysis of two different solution approaches are presented. This is followed by the results of the two methods.

2   Problem Description

The problem of locating all the splices can formally be described as follows. Consider the underlying (e.g., road or trench) network to be an undirected graph $G = (V_G, E_G)$ with non-negative edge weights. The vertices V_G and edges E_G represent the intersections and streets, respectively. Each $v \in V_G$ has coordinates (x_v, y_v) and d_{ij} denotes the weight (length) of an edge $(i,j) \in E_G$. Let $T = (V_T, E_T)$ with weights w_{kl} for $(k,l) \in E_T$ be a weighted tree representing a subset of the copper network.

From now on, T strictly refers to such a subset of the networks, starting at the cabinet. Let the root of T, denoted by s, be a cabinet while the leaves, $t_i, i = 1, \ldots, M$, coincide with the end of cables. The splices that need to be located are thus given by the inner nodes of T. Let $\mathcal{X}_G$ and $\mathcal{X}_T$ be defined as the infinite set of all (x, y) coordinates pertaining to the geographical location of the graph G and the copper network T, respectively. The aim is to find (x_u, y_u) such that $(x_u, y_u) \in \mathcal{X}_G$ for every u for $u \in T \backslash \{s, t_1, \ldots, t_M\}$. In other words, all nodes in T need to be located such that their coordinates correspond with either a node or a point on an edge in G. This is illustrated in Fig. 1. Here the goal is to find the coordinates of splices 1 and 2, from the tree T on the road graph G. Note that the copper lengths of the red lines on top of G, are of the same length as in T. The splices are assumed to lie on the red edges at the appropriate distance from the end cables and cabinet.

3   Pre-process

In this section some pre-processing and background work is covered to prepare the data to be used by the two methodologies.

In the remainder of this paper it is assumed that the structure of the copper network is known, which includes the following:

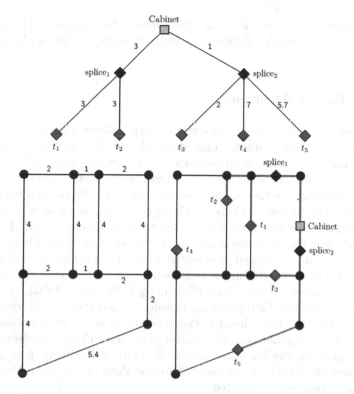

Fig. 1. Top: Copper network (T). Bottom left: Road network (G). Bottom right: Splice locations on top of road network. (Color figure online)

- The topology of the network, i.e., which nodes are connected to one another in the network;
- The network consists of a concentration point and then branches out (without cycles) to the end points;
- The network follows some known underlying street or trench pattern;
- The (estimated) copper-lengths between each connected pair of nodes in the network are available digitally;
- The (estimated) locations of the concentration point are available digitally, and;
- The (estimated) locations of the end of the cables are available digitally.

After establishing the geographical structure of the underlying network, known nodes of the copper network have to be added properly to G. For a given node $w \in V_T$ (either a leaf or the root of T) it is assumed to be connected to the graph of the road network at the nearest edge, say $e_w \in E_G$, based on their locations. For a more sophisticated approach, see [10].

Once the root of the copper tree is added to G, the size of the graph is reduced. This is done for the benefit of memory that is needed the keep track of the graph, as well as to improve the speed at which the remaining nodes are added to G, as there are now less edges that need to be evaluated. The reduction of the graph is done by determining the shortest path to every node in G from the root.

4 Path Based Approach

The first method aims to find the locations of the splices based on (s, t_i)-paths of given lengths on the graph G. This approach is based on the following idea. If there is only one possible path between s and t_i of the "correct" length, that being the length corresponding with the lengths in T, then it has to be the case that the splices between the cabinet and the end-cable lie on this path on the road network. The problem of locating the splices can now be seen as a series of path finding problems for paths of exact length. This search is limited to simple paths. A path q is called simple if nodes are visited at most once in q.

Formally, given a weighted undirected graph G, we aim to find a simple (s, t_i)-path of a given length, under the assumption that at least one such path exists. This is known as the Exact Path Length Problem (EPL). [11] gives a solution for solving the EPL problem. However, since the data is uncertain, it may very well be the case that the exact length is not always present in the graph. Instead, we use here the K-Shortest (Loopless) Paths Problem. First an overview is given of the literature on the K-Shortest Path problem. Next the specific implementation issues of the K-Shortest Path problem, for the problem under investigation, are elaborated.

4.1 Literature Review

Research regarding to the K-Shortest Path Problem in a network date back to as early as 1959 [7] and 1960 [2] who were amongst the first to propose methods for solving this problem. Yen [13] proposes an algorithm which, at the time of publication, had the best worst case complexity of $\mathcal{O}(KN^3)$ on the number of operations, where N is the number of nodes in the network. Yen's algorithm is limited to finding simple paths but does allow weights to be negative.

In a paper published in 1997, Eppstein [3] introduces a method, Eppstein's Algorithm (EA), to determine the K-shortest paths in a directed graphs where cycles are allowed. EA improves the worst case time complexity of Yen to $\mathcal{O}(|E| + N \log N + K)$. Jiménez and Marzal propose two separate variants on EA, Recursive Enumeration Algorithm (REA) [8] and Lazy Variant of Eppstein's Algorithm (LVEA) [9]. While only LVEA has the same worst case time complexity as EA, the authors show that both methods perform better than EA in practice. Next, [1] proposes another method based on the design of (LV)EA as an on-the-fly directed search heuristic. The algorithm, known as K^*, makes use of the so-called A^* path finding algorithm. A^* ([4]) is a directed search algorithm for the K-Shortest Path Problem which operates in many ways similar to that of Dijkstra. The difference arises when evaluating nodes. Where Dijkstra simply computes the length of the path from the source to the node in question (say $d(s, u)$), A^* uses a heuristic function, $f(s, u)$ which also includes an estimate for the remaining distance $h(u)$ in the path ($f(s, u) = d(s, u) + h(u)$).

Furthermore, Hershberger *et al.* [5] propose an algorithm for the K-Shortest Path Problem via a Path Replacement Problem [6]. This algorithm is particularly interesting because it was tested on geographic information system (GIS) road networks. This is particularly interesting as it is a similar application of the K-Shortest Path problem. [5] is directly compared to Yen's algorithm and is shown to perform better. The improvement, however, seems to increase with the size of the problem. They show that, for example, from Washington DC, to New York City, the network contains 6,439 nodes and 15,350 edges. The performance difference between the two algorithm, for generating 50 shortest paths, is not that large. The size of that problem is much larger than the problem in this paper. Yen's Algorithm is more efficient for small (low number of edges) problems and will be used as starting point in this paper.

4.2   Implementation

Since Yen's algorithm does not look for the paths of exact length explicitly, an adjustment is proposed here. Say there is an (s, t_i)-path P^* in G of length equal to the length of the copper cable between these two nodes. Given all possible paths from t_i to s there is some K' for which the K'th shortest path is P^*. Since K' is not known beforehand, paths will be found in ascending order until all paths of the length we are interested in are found. As Yen's algorithm terminates once all K shortest paths are found, the stop criteria in the implementation of the algorithm is altered such that the algorithm continues looking for paths until all paths smaller than some desired length are found. There are, however, some issues that arise when implementing this method.

Exploiting the copper tree structure. Often, a copper path shares splices with another path. Say that two paths share splices up to k, first the locations of the nodes in the first path are found. Then then path to k is reused. This is done by applying Yen's algorithm to find the path(s) from k to the second path's t of the length corresponding to the copper length between the two nodes.

Dealing with uncertainty. Assuming there is uncertainty in the coordinates and length, there may not be paths between nodes on the underlying network of a length coinciding with the data. Instead of having Yen's algorithm terminate when all paths of the desired length are found, it is used to find all paths of a length within a given interval, $[L_{lower}, L_{upper}]$. This is achieved by setting the stop criteria to be at the first path found with a length greater than L_{upper}.

Adding the missing splices. Once all the candidate paths between two nodes are identified, the algorithm proceeds to determine the location of the remaining splices on this path. The splices are placed based on their copper distances to a previous node, either by starting at t_i and working their way to s or the other way around. Given the uncertainty, there are three situations possible. Namely the path found is either longer than, shorter than, or equal to the path length the data implies. Issues arise at the first two scenarios. Now the nodes are placed based on their relative distance between t_i and s.

5 Bottom-Up Approach

As an alternative to the path based approach, an approach is presented that takes advantage of the copper network's tree structure more explicitly to construct a bottom-up algorithm. The name "bottom-up" is used because the algorithm starts at the leaf nodes of the copper tree and uses their locations and the structure of the copper tree to iteratively determine the locations of the nodes one level higher. When the locations of those nodes are determined, the process is repeated until all levels of the copper tree are evaluated. Under the assumption that the locations of the cabinet and end nodes, and copper lengths are indeed correct, a splice should be located such that there exists paths to all to its connected nodes of a length coinciding with the length of the copper cables between them. In order to apply the bottom up approach to find the location of common parents, a specific structure of the tree is assumed. Since T may have inner nodes with only one child, a reduced tree T' needs to be created in which these nodes are hidden.

Determining possible inner node locations. First possible splice locations are found based on all children individually and then is seen if there are any overlapping locations. It is at these points that the splices would possibly be located in reality. When estimating the splice locations, two methods are distinguished for creating paths on G. For the first method, all possible paths from t_i are enumerated of a desired length. This is done by starting at t_i and finding all points on G, for which there is a path between those point and t_i with a length equal to the length of the copper cables between t_i and the splice that is searched for. Secondly, paths are looked for in a method similar to a depth first search (DFS). Starting at a selected node (t_i) in G, the graph is explored until either a point in G is reached, such that the length of the path is equal to the length of the copper cable, or until the path can not be further extended. In the former case, the location of the point where the path ends is denoted as a possible splice location. All nodes along this path are marked as visited. Then, we proceed by backtracking in G until a node is reached that has a neighbour which has not yet been visited. From here, we attempt to find another path, which branches off at this point in the graph. A node cannot be revisited by a path after a branching point. This means that the DFS could potentially find fewer paths than when enumerating all paths, but does also exhaust the possibilities more quickly. This is the core difference between the DFS and enumerating all paths.

Choosing best splice location. Once all possible splice locations are determined, one of these locations has to be selected to place the splice at. Under the assumption that the data regarding the graph and locations are completely reliable, we first determine which possible locations are shared: these locations are labelled as parent by all the child nodes. In this case we need to pick the best one. This is done by taking the distance to the next node into consideration. The issue of locating the possible locations is further complicated when taking the uncertainty of the data into account. In this case, we do not have a guarantee that possible splice locations share the same coordinates. One way to take this into account is by redefining what we mean with overlapping locations. Instead of looking for points on the graph that have the same coordinates, we will determine the shortest path between pairs of possible splice locations and see if they are close enough together. If the distance is small enough, then we consider the two nodes to overlap and place the splice midway between the two points.

Finding non-reduced splices. Once we determined the expected location of the splice, we keep track of the paths that lead to the chosen solution. From here it is checked whether there are any splices in the full copper tree (as opposed to the reduced tree T') between t_i and the splice we originally sought after. If so, we follow the path and add the remaining splices based on the distances along the copper network.

6   Numerical Results

In this section, both methods are applied on a number of cases. Their performance on different scenarios is discussed.

Two sets of test cases were created. First two example cases of a copper network, XY and AM are presented, both are based on the road network in Amsterdam. These copper networks were created such that the locations of the splices are known, but the copper distances are not entirely correct. Secondly, we have a number of real world scenarios (CE, JE, GH and KF), based on scanned maps with the real information and estimated data from other sources. Table 1 shows the properties of these cases.

Table 1. Properties of the copper- and road network for our cases.

	XY	AM	CE	JE	GH	KF
Number of splices	34	75	48	65	8	43
Total copper length	1828	987	1305	1096	620	1084
$|V_R|$	246	133	208	63	30	85
$|E_R|$	320	163	247	65	29	95
$|V_G|$			416	358	238	316
$|E_G|$			501	392	276	347

The algorithms were applied on these test cases. The algorithm's performance is evaluated by looking at its run time and accuracy. When testing their accuracy, we look (1) at the number of splice locations that are determined and (2) the accuracy of the locations of the found splices. In the case of a known network, we keep track of which ones have been found and compare the different methods. We evaluate how many of the splices are found. In order to test the influence of the order in which we search for paths in the K-Shortest Path problem the paths are sorted based on their total length in ascending (KSP^{Short}) and descending (KSP^{Long}) order. We apply both methods on all of the cases. The mean squared error (MSE) is determined for each found splice location as a metric to compare the different methods. For each splice found, we take the squared distance between the coordinates found by the given algorithm and the actually (or estimated, based on map information) location of the splices.

When looking at the two variations of the K-Shortest Path problem approach for the cases XY and AM, we quickly notice that sorting the paths in descending order of length, gives the best result in terms of running time. This is likely due to the fact that the longer paths, more often than not also have more splices along the path. This means that more intermediate splices are found earlier on (Table 2).

Table 2. The results of cases XY and AM.

	Case XY			Case AM		
	Found	MSE	Time	Found	MSE	Time
KSP^{Short}	28/34	18.511	3.659	64/75	5.604	5.323
KSP^{Long}	28/34	17.719	31.113	62/75	12.088	16.582
BU^{dfs}	-	-	-	62/75	300.170	4.404

As for the Bottom-up approach, we see that for the case XY, it was unable to run appropriately. This is due to the nature in which it finds paths. Since not all possible solutions are evaluated in this method, it may occur that the possible paths created on the graph did not lead to any feasible splice locations. For the case AM, however, it was successful in completely running as intended. We see that it found nearly as many splices as the best performing K-Shortest Path problem while having the lowest running time. However, the location of the splices based on this method is clearly the least accurate.

Table 3. The results of cases GH and KF.

	Case GH			Case KF		
	Found	MSE	Time	Found	MSE	Time
KSP^{Short}	8/8	0.746	18.901	42/43	8.024	5.674
KSP^{Long}	8/8	0.723	2.983	43/43	132.346	4.617
BU^{dfs}	8/8	0.899	2.615	43/43	6044.500	7.485

We now look at how the methods perform on some real world cases. The first case GH is a relatively small example. We see in Table 3 that both methods find all splices and that KSP^{Long} is noticeably faster than KSP^{Short}. For BU^{dfs} we see similar results to the two KSP methods when looking at the splice locations.

For the second data set KF in Table 3 the level of inaccuracy of BU^{dfs} is noticeable. This method is unable to handle the inconsistencies between the given copper lengths and the actually path lengths found on the underlying network between the cabinets and end of cables. In Table 3 we see that KSP^{Long} finds more splices than KSP^{Short} but with a noticeably larger MSE. The higher mean error is attributed to the extra splice found and the reliability of its coordinates. This issue is one that lies beyond the algorithm, rather shows a clear example on the difficulty of the problem presented in this paper as a whole.

Table 4. The results of the cases JE and CE.

	Case JE			Case CE		
	Found	MSE	Time	Found	MSE	Time
KSP^{Short}	65/65	7.496	7.252	48/48	9.090	22.008
KSP^{Long}	60/65	10.679	6.989	46/48	13.934	36.619
BU^{dfs}	65/65	838.866	8.141	46/48	66.557	17.851

Table 4 shows that KSP^{Short} finds a few more splices with small differences than KSP^{Long} when looking at the accuracy and run time. Once more BU^{dfs} is the poorer performing method. While it is capable of finding an estimate for more splice locations than KSP^{Long}, these are clearly not as reliable as the MSE indicates.

The results for the final case can be seen in Table 4. The consistency in the difference in performance between BU^{dfs}, KSP^{Short} and KSP^{Long} holds again. While we did not test the Bottom-up approach with the variation of full enumeration nor on exact cases as we did with our modified Yen's algorithm, it is still clear that the Bottom-up approach has some difficulties with accuracy.

7   Summary and Conclusions

The goal of this paper was to develop two possible methods to reconstruct the physical path of underground networks, illustrated by finding the locations of splices in the copper network for telecommunication services. The first method discussed was a path based approach based on finding paths of given length in a network. Secondly, we introduced a so called bottom-up approach with the aim of taking a more direct advantage of the structure of the copper network.

Testing on a series of cases shows that these methods can indeed find all splices at their correct location when the input data contain no errors. However, the same accuracy cannot be guaranteed when expanding the cases to be more realistic and thus containing errors or uncertainty, but we can conclude that the path based approach is much better in handling this uncertainty. The testing showed also that the use of found splices in further iterations can greatly improve the algorithms performance.

In further research both methods could be improved by temporarily accepting multiple splice locations and testing which one may lead to better solutions. A variation of the problem presented in this paper, is the case where the paths do form cycles on the underlying network.

Acknowledgments. The authors like to thank Niels Neumann for his helpful comments and discussion.

References

1. Aljazzar, H., Leue, S.: K*: a heuristic search algorithm for finding the k shortest paths. Artif. Intell. **175**, 2129–2154 (2011). Elsevier
2. Bellman, R., Kalaba, R.: On kth best policies. J. Soc. Ind. Appl. Math. **8**(4), 582–588 (1960)
3. Eppstein, D.: Finding the k shortest paths. In: 1994 Proceedings of 35th Annual Symposium on Foundations of Computer Science, pp. 154–165. IEEE (1994)
4. Hart, P.E., Nilsson, N.J., Raphael, B.: A formal basis for the heuristic determination of minimum cost paths. IEEE Trans. Syst. Sci. Cybern. **4**(2), 100–107 (1968)
5. Hershberger, J., Maxel, M., Suri, S.: Finding the k shortest simple paths: a new algorithm and its implementation. ACM Trans. Algorithms (TALG) **3**(4), 45 (2007)
6. Hershberger, J., Suri, S.: Vickrey prices and shortest paths: what is an edge worth? In: Proceedings of 42nd IEEE Symposium on Foundations of Computer Science, pp. 252–259. IEEE (2001)
7. Hoffman, W., Pavley, R.: A method for the solution of the nth best path problem. J. ACM (JACM) **6**(4), 506–514 (1959)
8. Jiménez, V.M., Marzal, A.: Computing the K shortest paths: a new algorithm and an experimental comparison. In: Vitter, J.S., Zaroliagis, C.D. (eds.) WAE 1999. LNCS, vol. 1668, pp. 15–29. Springer, Heidelberg (1999). doi:10.1007/3-540-48318-7_4
9. Jiménez, V.M., Marzal, A.: A lazy version of Eppstein's K shortest paths algorithm. In: Jansen, K., Margraf, M., Mastrolilli, M., Rolim, J.D.P. (eds.) WEA 2003. LNCS, vol. 2647, pp. 179–191. Springer, Heidelberg (2003). doi:10.1007/3-540-44867-5_14
10. Neumann, N., Phillipson, F.: Connecting points to a set of line segments in infrastructure design problems. In: 21st European Conference on Networks and Optical Communications (NOC 2016), Lisbon, Portugal (2016)
11. Nykänen, M., Ukkonen, E.: The exact path length problem. J. Algorithms **42**(1), 41–53 (2002)
12. Phillipson, F.: Efficient algorithms for infrastructure networks: planning issues and economic impact. Ph.D. thesis, VU Amsterdam (2014)
13. Yen, J.Y.: Finding the k shortest loopless paths in a network. Manag. Sci. **17**(11), 712–716 (1971)

Design and Realization of Mobile Environmental Inspection and Monitoring Support System

Hyung-Jin Jeon[1(✉)], Seoung-Woo Son[1], Jeong-Ho Yoon[1], and Joo-Hyuk Park[2]

[1] Future Environmental Strategy Research Group,
Korea Environment Institute, Sejong, Korea
{hjjeon,swson,jhyoon}@kei.re.kr
[2] Solution Business Group, SundoSoft, Seoul, Korea
jhpark@sundosoft.co.kr

Abstract. If as mobile devices are introduced and cutting-edge information and communication technology (ICT) is developed, the need for mobile service at environmental inspection and monitoring work task sites in environmental contaminant discharging facilities is increasing. In this study, a hybrid app combining the mobile application technology and the mobile web technology was developed based on the environmental guidance and examination procedures, the analysis of the relevant laws and systems, and the results of the interview and questionnaire survey performed with relevant public officers. It uses Android and iOS which are the operating systems widely used in Korea. Considering the characteristics of the environmental guidance and examination procedures, the developed application provides information about the physical state or performance of environmental contaminant discharging facilities and the relevant authorities. To meet the required efficiency at the worksite and to provide optimized system functions, an actual application and operation test was performed by establishing a test-bed.

Keywords: Environmental inspection and monitoring · Mobile · Environmental contaminant discharging facility · Real-time information · Tablet PC · Environmental compliance

1 Introduction

The distribution rate of smartphones is increasing worldwide. As of March 2015, the average distribution rate of smartphones in 56 countries reached 60 %. The country with the highest distribution rate was UAE with 90.8 %, followed by Singapore (87.7 %), Saudi Arabia (86.1 %), South Korea (83.0 %), and Spain (82.8 %) [1]. Due to the wide distribution of reliable wireless network environment and mobile devices with improved mobility, the mobile office is drawing attention as an industrially competitive technology. The mobile office system is a product of the information technology in which 'smart' elements (the enablement of easy installation and utilization of various applications) are integrated with 'mobile' elements (rapid spatial change and speed of operation) [2]. Since the mobile office system enables intra-company and inter-company

© Springer International Publishing AG 2016
G. Fahrnberger et al. (Eds.): I4CS 2016, CCIS 648, pp. 59–71, 2016.
DOI: 10.1007/978-3-319-49466-1_5

communication as well as real-time information sharing and work process, the range of utilization is gradually expanding within public institutions to major or minor enterprises; expecting an increase in the efficiency of work [3]. The influence of this technological trend on the environment of public institutions is represented by 'smart work.' Workers are able to do their task with mobile devices regardless of time and place, breaking free from the conventional work pattern of using personal computers in designated offices. To promote the 'smart work,' the Defense Information Systems Agency (DISA) of the U.S. Department of Defense established a mobile working environment using laptop computers and virtual private network so that about 70 % of the workers in Washington D.C. may work away from offices. The British Police Office connects regional police stations to a central server, and a PDA (Personal Digital Assistant) is used to automatically upload or inquire criminal records as well as investigation, inspection, and arrest records. This allows for the redundant and tedious document works to be comprehensively done at the actual sites. Bearing Point Japan distributed smartphones for business use to the executives and consultants to provide business mail server connection. In Korea, smartphone-based mobile work environment is being established by communication enterprises such as KT and SKT, and major enterprises such as Samsung and Posco [4]. Recent investigation have been focused on using mobile technology to reduce the efforts involved in generating inspection documentation and to simplify the customer's business model [e.g., 5–9]. To scientifically counteract the act of environmental contaminant discharging, which is recently becoming more sophisticated and subtle, a comprehensive information system regarding facilities that discharge environmental contaminant needs to be established. It will enhance the citizens' accessibility to environmental information so that they may search the information about the discharging facilities and immediately report the act of illegal environmental contaminant discharging using smartphones, based on mobile positioning technology. The businesses that discharge wastewater and air pollutants should obtain permission for the installation of discharging facilities by undergoing strict preview and inspection. However, in cases when the discharging facilities or pollution prevention facilities are not operated normally by accident or on purpose in the operation process, contaminants may be discharged to public water, air or soil, resulting in pollution of the environment and damage to the public health. Therefore, the public officers of the state government or local governments visit these places of business to guide and examine the discharging facilities or pollution prevention facilities, inspecting whether the facilities are appropriately operated or not [3]. In this study, an environment monitoring task supporter system was established to use contaminant discharging information and measurement data from business locations for environmental inspection and monitoring. In addition, the position-based technology of mobile devices was used to improve procedures and increase efficiency as means of further enhancement of the environmental inspection and monitoring work.

2 Background

The Ministry of Environment, the National Institute of Environmental Research, the Korea Environment Corporation, and local governments are separately managing huge amount of information and database related to the business places all over Korea, the current status of contaminant discharge, and the environmental quality, including the 'Korea Pollution Sources Data,' the TMS (data of air quality and water quality related business places), the real-time 'Allbaro System' (managing the generation and treatment of designated waste including medical waste), and the air pollution measurement network data. Currently, however, the data are not used effectively in the environmental inspection and monitoring works. In this study, the current status of the works implemented by the Ministry of Environment and the local governments was investigated to evaluate the environmental inspection and monitoring works for individual business locations. In addition, to identify the current problems of the environmental inspection and monitoring works related to contaminant discharging businesses, and to present methods of improvement, a current status survey was performed through questionnaire survey and interviews with the relevant public officers. On the basis of the current status survey results, an environment monitoring task supporter system was developed for the advancement of the environmental inspection and monitoring. A test bed was also established and operated to test the system applicability on the sharing of contaminant discharging information with relevant authorities, integration of geographical information function, and mobile environmental inspection and monitoring.

2.1 Current Status of Inspection and Monitoring Works

To increase the efficiency of the environmental contaminant discharging facility inspection and monitoring works, and to promote the normal and appropriate operation of discharging facilities and pollution prevention facilities, individual local governments implement integrated inspection and monitoring works according to the Regulations on Integrated Inspection and Monitoring and Examination of Environmental Contaminant Discharging Facilities, Etc. The inspection and monitoring works include scheduled inspection and monitoring, and nonscheduled inspection and monitoring. The Environmental Monitoring Groups of the Ministry of Environment and the River Environmental Management Offices implement special guidance and planned clampdown. The planned clampdown is supposed to be implemented by the collaboration of the local governments, the Environmental Monitoring Group, and the Prosecutors.

The main works of the Environmental Monitoring Groups of the Ministry of Environment are planning and implementing of special monitoring against the act of environmental polluting, integration and coordination of monitoring and controlling works relevant to contaminant discharging facilities, the supervision and support of the contaminant-discharging facilities-inspection and monitoring-works implemented by local governments, auditing of local governments in collaboration with the state government, the improvement and operation of environmental crime controlling and investigation systems, and the operation of Environmental Special Judicial Police system. The organizations of local governments in charge of the environment manage the inspection and monitoring. The environmental works managed by the local governments include the permission, report, inspection and monitoring, and administrative disposition with respect to all the contaminant discharging businesses including air and water pollutant discharging facilities, sewage and livestock wastewater discharging facilities, and toxic substance registration companies. Two ways to approach monitoring environmental contaminant discharging facilities are either by classified as medium-specific regulation or integrated, cross media permitting. Korea has adopted the medium-specific regulation approach, similar to the approach employed by the U.S., Japan, Russia, and China. Finland, France, Netherlands, and U.K. have adopted the integrated, cross media permitting approach [10]. National enforcement databases like the U.S. Environmental Protection Agency (EPA) enforcement and compliance history online database (ECHO) do suggest that monitoring and enforcement intensity under any given statute at any given point in time varies quite signification across states. However, recordkeeping, data management, industrial composition, facility characteristics, and many other factor at least partially explain enforcement across states [11]. The targets of the inspection and monitoring are varied depending on the opening and closing of business places or facilities. As the end of 2014, the number of target business places was 64, 254 (Table 1). Amount of pollutants can be classified by five types of discharging facility. Type 1 of business place is for annual amount of 80 tons or more, Type 2 for 20 tons to 80 tons, Type 3 for 10 tons to 20 tons, Type 4 for 2 tons to 10 tons, and Type 5 for less than 2 tons.

Table 1. Current status of environmental emission facility (2014)

Classification	Type of discharging facility					
	Type 1	Type 2	Type 3	Type 4	Type 5	SUM
Total emission	2,547	2,356	3,693	12,600	46,058	64,254
Air pollutants	567	444	596	3,115	5,417	10,139
Waste water	218	407	690	839	7,991	10,145

2.2 Current Status of Inspection and Monitoring Works

The violation rate found by the inspection and monitoring implemented by the Ministry of Environment and local governments from 2012 to 2014 was compared, and the result showed that the violation rate found by the Ministry of Environment was three times higher than that of the local governments (Table 2). The reason why the violation rate found by the local governments was significantly lower was because most of the heads of local governments were more interest in stimulating local economy than in preventing environmental pollution or conserving environment through environmental regulations. Thus neglecting the inspection and monitoring works. In addition, since the public officials in charge of the inspection and monitoring works participate in other types of works as well, they do not have expert knowledge. Their enforcement is often perfunctory. Moreover, while the contaminant discharging done by the discharging companies is becoming more sophisticated and subtle, the organization, human resources, and equipment of the local governments for the inspection and monitoring are not sufficiently aided to successfully carry out monitoring missions.

Table 2. Comparison of the violation rate found by the environmental inspection and monitoring between the ministry of environment and local government

Classification	2012	2013	2014
Ministry of Environment	26.5 %	28.9 %	28.0 %
Local Government	6.1 %	7.8 %	8.6 %

Violation rate: The ratio of the incidents of violating the environment related acts among the number of actual examination works with respect to the business places in the relevant year

2.3 Current Status of Inspection and Monitoring Works

Interviews were carried out with the relevant public officials who were in charge of the inspection and monitoring works for the contaminant discharging businesses in the Ministry of Environment and local governments. The River Environmental Management Offices of the Ministry of Environment were visited two times on June 26 and 27, 2014, and the local governments were visited two times on August 7 and 8, 2014 for the interviews [12]. During the interviews, questions were asked about the overall difficulties and requirements of the inspection and monitoring works and the discharging information management as well as establishment of a new system reflecting the solutions to the problems (Table 3). A questionnaire was prepared based on the interview materials; a survey was performed with the public officers who were in charge of the inspection and monitoring works all over Korea. The survey was performed as an internet survey (Google Docs) by sending an e-mail letter to the individual public officers who were in charge of the inspection and monitoring works. The responses from 137 public officers who sincerely responded to the questionnaire were used for the analysis.

Table 3. Content of the interviews performed with the relevant public officers of the ministry of environment and local government

Interviews	Content
Regional Government	Without the use of the present information system, use separate form
	Data management through information system
	Build a system for data share and communication
	Licensing, inspection and guidance process will be linked by information system
	Basic data must be fully reflected in the system
	Information provided through a information system
	The need for integrated traceability
Local Government	Utilizing its own system for traceability
	Workload by new information system
	Systems for information sharing and communication
	Without the use of the present information system, use separate form
	The need for location information and statistics service

Although the inspection and monitoring works require expertise knowledge, most of the public officers in charge of the works had a career less than 5 years, 27 % of them having a career less than 1 year, indicating that the human resources had a low level of expertise and monitoring capability. A question was asked to investigate if just one public officer managed the works for the air quality and water quality in an integrated manner or different officers took charge of different media. The result showed that 50 public officers (36 %) managed the environmental management works in an integrated manner and 87 public officers (64 %) managed the works in each media separately. Some difficulties experienced by the officers were the complicated administrative procedure caused by the complex systems and binary working systems, and the difficult data management due to the separated information systems. As methods of resolving these difficulties, the public officers required the establishment of an environmental monitoring task supporter system and sharing of discharging information with relevant authorities. This study was conducted by reflecting these requirements.

3   Approach

The results of the interviews, survey, and the work process of the environmental inspection and monitoring works were analyzed. The analysis showed that a system enabling to inquire and utilize the information about the current status of discharging facilities, contaminant measurement results, and violation history should be prepared for the inspection and monitoring works. Also, the work system should be improved to enhance the expertise and efficiency of the environmental inspection and monitoring works. To achieve this goal, in this study, an environmental monitoring task supporting service platform was established by interconnecting information about the discharging businesses and the relevant authority information. In addition, a system to apply this platform to the actual works was prepared. Considering the characteristics of the environmental inspection and monitoring works, which is focused on the actual work sites, a mobile environmental monitoring task supporter system was developed. The actual application and operation test of the system is currently performed by establishing a test-bed.

3.1   Establishment and Design of Mobile Environmental Inspection and Monitoring Support System

The mobile environmental monitoring task support system was designed to enhance the efficiency of the administrative works by rationalizing the means and standards of the environmental policies, and also to interconnect the procedures of information input, inquiry, and public opening according to the work process.

It is necessary to establish a database with regard to the contaminant discharging business places in order to design and develop the mobile environmental inspection and monitoring support system. Daejeon and Gwangju were chosen as the test-beds to evaluate the system. The databases on air pollution and wastewater discharge were established by obtaining information about facility status, instruction, examination and administrative disposition history from the local administrative systems. In addition, the information was shared databases of relevant authorities for the integrated environmental information management (Fig. 1).

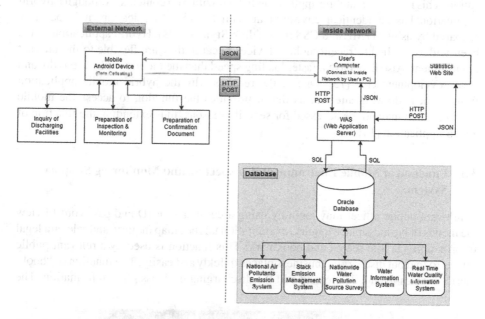

Fig. 1. The workflow of structural environmental inspection and monitoring support system

3.2 Implementation of Mobile Environmental Inspection and Monitoring Support System

The mobile environmental inspection and monitoring support system was developed as a hybrid application including functions of a mobile application and web operated under Android and iOS. The system was implemented after being analyzed and designed in the Electronic Government Framework 2.0 based on HTML (Hyper Text Markup Language). This enables the system with the utilization of flash files, video clips, and multimedia files without the limitations of operating systems or platforms (Table 4).

Table 4. Content of environmental inspection and monitoring information database

Content	Android	iOS
OS	Android 4.0	iOS6.0.1
Language	JAVA	Object C
WAS	RESIN	RESIN
DB (Server)	ORACLE	ORACLE
Tool	Eclipse	X-CODE

A hybrid application which has advantages of both mobile and web, is developed by preparing a mobile web confirming to web standards and interworking with an application. This results in a mobile application format with web contents. Although a hybrid application has an identical environment with a mobile application, it is partially prepared by using HTML and CSS (Cascading Style Sheets). Hybrid applications have been widely applied to recent mobile devices because they are flexible to the internet environment. Also they have a faster loading speed than the mobile applications do, and save development cost [13]. One of the reasons why the hybrid mobile application framework is drawing attention is that it provides the function to access the mobile resources that may not be accessed for security reasons, thus extending the function of a web-application [14].

3.3   Function of Mobile Environmental Inspection and Monitoring Support System

A relevant public officer may login by using a certificate or ID and password to view the menus of business place inquiry, examination targets, map inquiry, and relevant legal information. (1) Site search and bookmark: This function is used by a relevant public officer to find out discharging business places quickly and easily. The function of "bookmark" enables to add, delete, and edit the discharging business place information. The

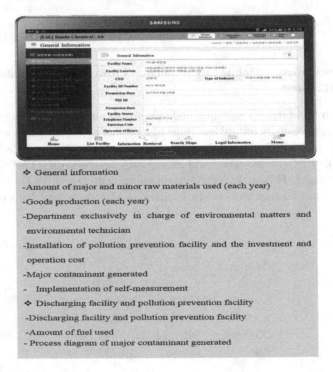

Fig. 2. Display screen and contents of location information inquiry and surrounding environmental measurement inquiry of environmental inspection and monitoring support system

main menus including the current status information, inspection history, confirmation report, map, and bookmarks may be used by searching the discharging business places (Fig. 1).

The current status menu of discharging business places provides general current status information including general information, amount of major and minor material used (each year), production goods, major contaminant generated, and self-measurement implementation information as well as the information about the discharging facility and pollution prevention facility in each business place. The discharging facilities are classified in the menu as wastewater discharging facility and air pollutant discharging facility according to the materials discharged from the discharging business places (Fig. 2).

(2) Location information inquiry: This function enables to inquire the positional information of the discharging business places which are the targets of the environmental inspection and monitoring works. As an additional function, the system provides the navigation function and enables to inquire the environmental measurement information of the surrounding regions (Fig. 3).

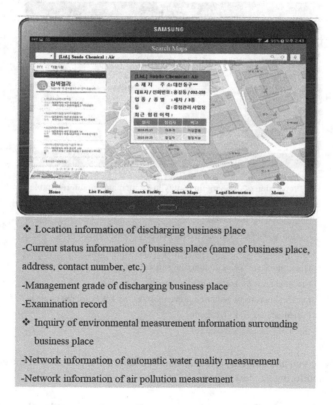

❖ Location information of discharging business place

-Current status information of business place (name of business place, address, contact number, etc.)

-Management grade of discharging business place

-Examination record

❖ Inquiry of environmental measurement information surrounding business place

-Network information of automatic water quality measurement

-Network information of air pollution measurement

Fig. 3. Display screen and contents of application for preparation of inspection and monitoring table of environmental inspection and monitoring support system

(3) Preparation of inspection and monitoring table: The discharging facility inspection and monitoring table may be prepared through the mobile system. The system provides the input format for brief examination results including the basic business information, environmental technician information, and discharging facility information as well as specific information for examination including the three-month operation status of pollution prevention facility, discharging facility modification information, and examination result information. The persons in charge of inspection and monitoring works should verify the main examination information including the discharging facility and pollution prevention facility information and report and administrative disposition information. The input of examination and its result for the mobile system is classified into permission and report of the discharging facility, operation and management, pollution prevention facility, self-measurement status, environmental technician working status, and report and administrative disposition status. The relevant officer may choose and input whether 'examination finished' or 'examination unfinished' as a result of inspection (Fig. 4).

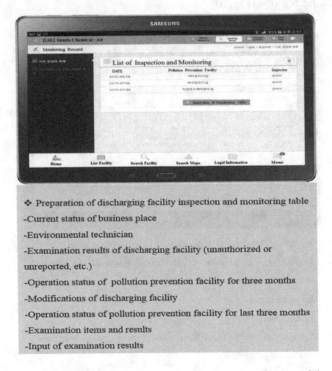

❖ Preparation of discharging facility inspection and monitoring table

-Current status of business place

-Environmental technician

-Examination results of discharging facility (unauthorized or unreported, etc.)

-Operation status of pollution prevention facility for three months

-Modifications of discharging facility

-Operation status of pollution prevention facility for last three months

-Examination items and results

-Input of examination results

Fig. 4. Display screen and contents of confirmation document application of the environmental inspection and monitoring support system

(4) Preparation of confirmation document: The system enables the preparation of confirmation documents including sampling confirmation document and violation confirmation document. The sampling confirmation document is required when an environmental inspection and monitoring officer takes samples from a business

place, and entrusts the samples to an inspection agency for the pollutant test. The violation confirmation document is required for violated items on the basis of the environmental inspection and monitoring results. The violation confirmation document should be issued for the violation of the legal regulations on the basis of the inspection and monitoring results according to Article 14 of the Regulation of the Integrated Inspection and Monitoring for Environmental Contaminant Discharging Facility, etc. The confirmation document should be written according to five W's and one H principle. When preparing either inspection and monitoring table, sampling confirmation document or violation confirmation document, the officer in charge of the inspection and monitoring should write the personal information of all those who participated in the examination at the end of the document and issue a copy of the document to business operators. The mobile environmental inspection and monitoring support system also implemented the signature function because those who examinate as well as those who receive the examination should sign on either inspection and monitoring table, sampling confirmation document or violation confirmation document prepared by using the mobile system (Fig. 5).

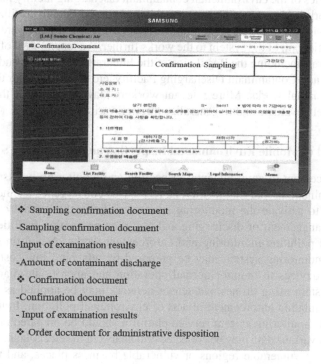

❖ Sampling confirmation document

-Sampling confirmation document

-Input of examination results

-Amount of contaminant discharge

❖ Confirmation document

-Confirmation document

- Input of examination results

❖ Order document for administrative disposition

Fig. 5. Display screen and contents of confirmation document application of the environmental inspection and monitoring support system

3.4 Test-Bed of Mobile Environmental Inspection and Monitoring System

The mobile environmental monitoring work supporting system developed in this study is currently applied to the actual work sites to improve the system by collecting the reviews from relevant public officers. The test-bed is used by ten relevant public officers of Daejeon and Gwangju in their on-site works. The effect of the mobile environmental monitoring work supporting system was analyzed by information technology effect model suggested by Delone and McLean (2014), and its analytical result indicated that the work productivity had been increased by 30 to 50 % in the aspects of the economic effect, safety and operation effect, and organization and culture effect [15].

4 Conclusion and Future Work

4.1 Test-Bed of Mobile Environmental Inspection and Monitoring System

In this study, a mobile environmental monitoring work supporting system was developed for characteristics of the environmental contaminant discharging facility inspection and monitoring works. Also, a test-bed was established for the effective utilization of the system. The opinions of the relevant public officers using the system were collected to extend the application of the system to the works further. An amendment of the relevant laws and systems, including the Regulation of the Integrated Inspection and Monitoring for Environmental Contaminant Discharging Facility, etc., is required for the system to be applied to actual works. More relevant workers may use and test the system to comprehensively evaluate and to generalize through various analyses.

4.2 Test-Bed of Mobile Environmental Inspection and Monitoring System

For the advancement of environmental inspection and monitoring works for environmental contaminant discharging facility, an ICT-based convergence technology should be developed to activate the monitoring system verifying the normal operation and appropriate management of discharging and pollution prevention facilities as well as environmental pollution monitoring and enforcement works. A real-time discharged contaminant monitoring system may be prepared by attaching low-cost 'internet of things'-based devices to monitor normal operation and appropriate management. A monitoring system using drones and sensor network may be applied as a method of performing regulation enforcement of acts of environment pollution. In the future, an environmental monitoring system service platform should be prepared and positively applied so that various data may be collected and analyzed to select planned crackdown targets including vulnerable regions or vulnerable business places, and to utilize the pollution measurement data of individual business places.

Acknowledgments. Many thanks to Yoon-Shin Kim from the University of Konkuk for proofreading this paper! This research was supported by the Korea Environmental Industry & Technology Institute (No. 2014001610002) and was conducted by Korea Environment Institute.

References

1. You, D., Ko, K., Maeng, S., Jin, G.: Mobile cloud service platform for supporting business tasks. J. Korea Inst. Inf. Commun. Eng. **17**(9), 2113–2120 (2013). https://goo.gl/rHnGLK
2. Nuckols, J.R., Ward, M.H., Jarup, L.: Using geographic information systems for exposure assessment in environmental epidemiology studies. Environ. Health Perspect. **112**(9), 1007–1015 (2004). https://goo.gl/VrWpYh
3. Hwang, G.: Environmental inspection and control - present state and policy direction. Korea Environ. Law Assoc. **37**(2), 75–101 (2015). https://goo.gl/PIZV9b
4. Lee, D., Yang, I.: Global cases of smart-work promotion. Korea Inf. Process. Soc. Rev. **18**(2), 90–99 (2011). https://goo.gl/KkpbkS
5. Asbahan, R., DiGirolamo, P.: Value of tablet computers in transportation construction inspection: ongoing case study of project in Pennsylvania. Transp. Res. Rec. J. **2268**, 12–17 (2012). https://goo.gl/sb8kCu
6. Nguyen, L., Koufakou, A., Mitchell, C.: A smart mobile app for site inspection and documentation. In: Proceedings of ICSC15 - The Canadian Society for Civil Engineering 5th International/11th Construction Specialty Conference, University of British Columbia, Vancouver, Canada (2015). https://goo.gl/EutIGn
7. Boddy, S., Rezgui, Y., Cooper, G., Wetherill, M.: Computer integrated construction: a 18 review and proposals for future direction. Adv. Eng. Softw. **38**(10), 677–687 (2007). https://goo.gl/llcFDQ
8. Kim, Y., Kim, S.: Cost analysis of information technology-assisted quality inspection using 21 activity-based costing. Constr. Manag. Econ. **29**, 163–172 (2011). https://goo.gl/L7otHR
9. Kimoto, K., Endo, K., Iwashita, S., Fujiwara, M.: The application of PDA as mobile 23 computing system on construction management. Autom. Constr. **14**(4), 500–511 (2005). https://goo.gl/9xHw2o
10. Organisation for Economic Co-operation and Development: Ensuring environmental compliance: trends and good practices (2009). https://goo.gl/EHO9Dh
11. Shimshack, J.: The economics of environmental monitoring and enforcement. A review. Annu. Rev. Resour. Econ. (2014). https://goo.gl/8yPTGL
12. Son, S., Yoon, J., Jeon, H., Myung, N.: Research on the improvement measures on the guidance and inspection for an environmental pollutant discharging company by using the problem analysis. J. Korea Acad. Coop. Soc. **16**(10), 6466–6474 (2015). https://goo.gl/KdB8xd
13. Choi, J., Kim, S., Kim, C., Lee, C., Joo, Y.: Development and implementation of mobile app for marine pollution responder. **19**(4), 352–358 (2013). https://goo.gl/1h5X77
14. Jung, W., Oh, J., Yoon, D.: Design and implementation of hybrid mobile APP framework. J. Korea Inst. Inf. Commun. Eng. **16**(9), 1990–1996 (2012). https://goo.gl/KwsTIG
15. Delone, W.H., McLean, E.R.: The DeLone and McLean model of information systems success: a ten-year update. J. Manag. Inf. Syst. **19**(4), 9–30 (2003). https://goo.gl/tDPk8v

Coding and Security

Re-visited: On the Value of Purely Software-Based Code Attestation for Embedded Devices

Maximilian Zeiser[✉] and Dirk Westhoff

Institut Für verlässliche Embedded Systems und Kommunikationselektronik,
Hochschule Offenburg, Offenburg, Germany
{Maximilian.Zeiser,Dirk.Westhoff}@hs-offenburg.de
http://ivesk.hs-offenburg.de/

Abstract. Remote code attestation protocols are an essential building block to offer a reasonable system security for wireless embedded devices. In the work at hand we investigate in detail the trustability of a purely software-based remote code attestation based inference mechanism over the wireless when e.g. running the prominent protocol derivate SoftWare-based ATTestation for Embedded Devices (SWATT). Besides the disclosure of pitfalls of such a protocol class we also point out good parameter choices which allow at least a meaningful plausibility check with a balanced false positive and false negative ratio.

Keywords: Remote code attestation · Wireless embedded devices · Time-bounded challenge response protocol

1 Introduction

For a trustworthy code execution on wireless embedded devices like sensor nodes, smart meters or other embedded IoT devices (at least) three functionalities have to be supported: a secured over-the-air (OTA) programming (Stecklina et al. 2015 [10]), a remote code attestation (RCA), as well as control flow integrity (CFI) (Stecklina et al. 2015 [11]). Whereas OTA-programming allows to remotely update respectively configure a node's code image from time to time, RCA is a protocol class to remotely verify or at least validate if the currently executed code image is still the originally OTA-programmed one. It is the objective of a CFI enriched executable to prevent code sequences from being re-shuffled such that a trustworthy code gets re-arranged to a malicious one. We state that CFI and RCA are both mandatory pre-requisites with respect to system security for embedded wireless devices and emphasize that both concepts are complementary to each other. CFI prevents the subversion of machine-code by integrating runtime checks into a binary such that the code's control flow fits to a given control flow graph. On the contrary, RCA validates that at *a given point in time* the code currently running on a node is still the one originally uploaded on it. Particularly this means that a challenge-response based RCA has to ensure that an

© Springer International Publishing AG 2016
G. Fahrnberger et al. (Eds.): I4CS 2016, CCIS 648, pp. 75–89, 2016.
DOI: 10.1007/978-3-319-49466-1_6

attacker is not able to store the bogus code in addition to the originally uploaded one and uses the originally uploaded code image solely for computing a challenge e.g. from a base-station resp. verifier when requested. The pre-dominant way to gain free memory space for its exploit is to compress the originally uploaded code image and only decompress if needed for a routinely incoming challenge-response run. Thus, whereas CFI aims that instructions from the originally uploaded code image cannot be re-shuffled in a meaningful but bogus way to build an exploit at runtime, RCA aims that free memory cannot be used in an undetected way for uploading the exploit at a later point in time once the sensor nodes have been deployed. Moreover, since RCA has never been designed to detect abnormal control-flow-modifications, some form of CFI protection would always be required in addition to the deployment of an RCA approach when aiming for an enhanced level of system security. However, we argue that vice versa also CFI benefits from RCA: Without RCA an attacker can e.g. upload besides the CFI enriched code image the same code image again, this time however without rewriting the code image's compiled binary. Basically, in Kumar, Kohler, and Srivastava (2007) CFI is preserved by maintaining a safe stack that stores return addresses in a protected memory region. Thus, in the remainder of this work we always assume a CFI enriched code image generated with the required preprocessing steps. Surely, our subsequent RCA analysis of software-based RCA also holds without a CFI enriched code image, but for the aforementioned reasons we argue that one should always apply RCA and CFI at the same time. The work at hand is structured as follows: Sect. 2 gives an overview of methods for code attestation with trusted modules. Whereas an introduction to purely software-based remote code attestation is given in Sect. 3. We illustrate and introduce the threat model and attacker model in Sect. 4 as well as basic concepts of time-bounded purely software-based RCA algorithms. A deeper explanation and conditions derived from the domain parameters are given in Sect. 6. In Sect. 7 we present our test implementation and derive mandatory configuration as well as set-up recommendations in Sect. 8. From our gained experiences we give recommendations in Sect. 9. Finally, we conclude our research in Sect. 10. Table 1 lists abbreviations which will be used in the remainder of this work.

2   Code Attestation with Trusted Modules

Hardware-based solutions like Trusted Platform Module (TPM) [6] based ones suit as a trust anchor for an authentic reporting of a system's software state. For embedded devices like mobile phones a Mobile trusted Module (MTM) [5] may serve as such trust anchor. However, since for more restricted embedded devices like sensor nodes such hardware-based modules may still turn out to be too costly for many application scenarios, purely software-based attestation is still an active research field regarding such category of device classes. A 'hybrid' solution sketch has been proposed by Sadeghi and co-authors when introducing to apply physically unclonable functions (PUF) [8]. Their approach combines software attestation with device specific hardware functions with the objective to

Table 1. Abbreviations used in this work.

Symbol	Meaning
RCA	remote code attestation
CI	code image
$C(x)$	compression of x
PRW	pseudo random word
F_{Total}	total size of the internal flash storage
$F_{BogusCodeImage}$	size of the bogus code image
P	cycle count of the SWATT algorithm
t	time threshold for a valid RCA response
t_t	roundtrip time between two nodes
t_c	time needed for the challenge computation of the RCA response
t_Δ	Extra time added to t, to reduce the false negative ratio
t_{if}	time overhead caused by the additional if statement

also detect hardware attacks. Although, considering PUFs is surely a promising building block for code attestation approaches for sensor nodes, a more detailed view on PUF systems shows that they require helper data adding time and area overhead to the system [3]. Helper data again may open an attack vector to reverse-engineer the PUF's pseudorandom outcome. Moreover, and even more important with respect to a practical and still secure RCA solution, in addition to the hardware PUF at the embedded device, a PUF emulator based on a mathematical model of the PUF (Arbiter PUF e.g. Ring Oscillator PUF) needs to be available at the verifier side. Obviously, such a code is not trivial to build by still reaching the envisioned security level of the hardware PUF. All these aspects motivated us to re-visit purely software-based remote code attestation protocols. The most prominent one from this class of protocols is SWATT which we describe at next.

3   Software-Based Code Attestation

The difficulty in purely software-based code attestation protocols lies within the fact, that the attacker has access to the same resources as the valid node, thus traditional secret-based challenge-response will not work. Instead, the physical constrains of the hardware and side-channel information are used. The main idea of such a protocol class is that an attacker requires additional effort for the response computation. As long as the additional time difference due to the additional effort and the limited resources of an embedded device are big enough, an attack can be detected. A typical time-bounded challenge-response protocol consists of three main parts, (i) a pseudo-random address generator, (ii) a read function and (iii) a checksum. At the prover and verifier side the generator receives the nonce as input and generates addresses, which are read from the flash and used as input to the checksum. This whole procedure is repeated P

times. These three functions and environmental conditions derive the concrete setting of parameter P (see Fig. 2). Moreover, compressed code images, like proposed in [12], may be used to prevent compression attacks. But to preserve the performance a purely hardware-based decompression would be necessary. However, compression attacks seem to have a lower impact than originally expected (Castelluccia et al. 2009 [2]) and seem to even be ignored in other works [7]. Therefore, we focus our current work on the evaluation of purely software-based remote code attestation, like proposed in Armknecht'13 [1]. We apply the framework view of their work and focus on the practical impact as well as the setting of the parameters, like the impact of the network based jitter or the practical requirements of the cycle count P. Furthermore we try to minimize the false negative rate, without allowing false positives. Soft Ware-based ATTestation by Sehadri [9] et al. as the most prominent candidate of purely software-based RCA is based on the timing of an embedded devices response to identify a compromised node. The prover has to build a check-sum of memory addresses chosen by the base station in a pseudo random manner. If a compromised embedded device aims to compute a valid response, it has to check if the chosen address contains malicious code. If this condition is true, the original bytes have to be either guessed or calculated. Either way additional time Δt is needed to compute a valid response (see Fig. 1). If Δt is big enough (we will later argue what this means) a compromised node can be distinguished from a valid node. However, the probability of a compromised node passing the code attestation without being detected, is still $Pr(\frac{F_{Total}-F_{BogusCodeImage}}{F_{Total}})^P$. In this equation $F_{BogusCodeImage}$ denotes the size of the bogus code image, F_{Total} denotes the size of the memory and P is the number cycles of the SWATT-algorithm. For a proper configuration of such remote code attestation protocol and depending on the assumed minimal size of a possible bogus code image as well as the total size of the memory a reasonable P has to be chosen by the administrator at configuration time.

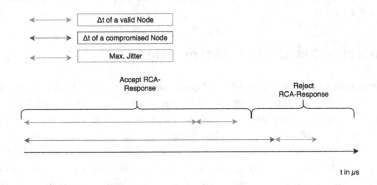

Fig. 1. Optimal case for the distribution of t.

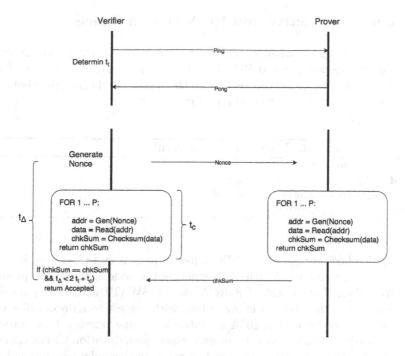

Fig. 2. Basic SWATT Model with initialization phase and attestation phase.

4   Threat and Attacker Model

We assume an attacker to gain physical access to the embedded device such that she can manipulate the content of the flash drive. Our definition of security is met, as long as an attacker is not able to execute arbitrary code or manipulate the CI without failing the next code attestation. Although the attacker gains physical access to the device we stress that all attacker models for purely software-based RCAs either implicitly or explicitly assume that the attacker will neither

(i) be able to tune the hardware clock,
(ii) equip the embedded device with a more powerful radio transceiver, nor
(iii) outsource computations to a more powerful external device.

All such approaches would allow the attacker to reduce the response time of the challenge-response protocol potentially answering within a time frame which was purely supposed to be possible for only uncompromised embedded devices. In case (ii) the attacker saves time in a multi-hop scenario due to faster transmission of the response whereas in cases (i) and (iii) the computation of the prover itself is tuned.

5   Security Objective and RCA Design Space

We want to investigate what we can at best expect from a purely software-based challenge-response based RCA (CR-RCA) approach like e.g. SWATT. A naive verifier-side inference with respect to the question whether the challenged wireless embedded device is compromised could be:

Algorithm 1. Naive RCA inference at the verifier.

1: **if** *response_time* < *threshold* **then**
2:    **if** *RCA_response* == *true* **then**
3:        **return** *CodeImage_unmodified*
4: **return** *CodeImage_modified*

In practice however, with a CR-RCA like approach and the above inference scheme we can not guarantee such easily whether the node has been compromised or not. We will see that at best a CR-RCA like e.g. SWATT achieves a *plausibility check* according to the above decision. Thus, with respect to a deeper discussion of the design space for an CR-RCA we introduce some relevant time intervals. With t_t we do denote the *arithmetic mean* transmission duration for the challenge as well as the response message over the radio in a single-hop and noiseless (without jamming) CR-RCA setting. With t_c we denote the *arithmetic mean* for the computation time for the response at a challenged and uncompromised node such that a reasonable approximation for the response time $t_{uncompromised}$ is $t_{uncompromised} \approx 2 \cdot t_t + t_c$. It is obvious that the confidence intervals for both arithmetic mean values t_t and t_c are subject to variations whereas the variation for the first value is typically larger. One can state that the overall magnitude of the confidence interval for the arithmetic mean value t_t mainly depends on the single-hop radio distance between sender and receiver. However, once nodes are deployed and their single-hop distance remains constant, the magnitude of t_t's confidence interval is marginal (however not neglectable) as pointed out by Elson and co-authors [4]. The magnitude of the confidence interval for t_c also tends to be small. In fact, t_c itself largely varies with respect to the concretely chosen percentage of addresses of the memory which are considered to be hashed together with the nonce. However, once this configuration decision for the CR-RCA is taken, one can argue that the confidence interval's variation of t_c is again marginal. This may hold under the precondition that the task for computing the response is *always* scheduled with the highest priority. Nevertheless, due to the above uncertainties when calculating $t_{uncompromised}$, a setting *threshold* := $t_{uncompromised}$ will always result in a considerable percentage of false positives and false negatives. Moreover, when the size of the memory F and the number of chosen addresses P are equal (resulting in a most probably too extensive calculation duration t_c) no additional attack possibility to feed in bogus code is given. Vice versa, with a more practical calculation duration t_c resulting from a smaller ratio $\frac{F}{P}$, the risk for undetected bogus code is again increasing such that even when the response $RCA\_response == true$ is calculated at the verifier within the duration $t_{uncompromised}$, there may still be a marginal probability for a compromised node passing the CR-RCA run. One way to deal with false negatives is to add some carefully balanced time t_Δ such that $threshold :=$ $t_{uncompromised} + t_\Delta$. This may avoid too many false negatives decisions with the consequence to eventually initiate another required code image update process.

Remark: One could argue that an attacker could also jam the radio. This surely increases t_t. Be aware that such an attack would not have the same intention of an adversary who yet uploaded a bogus code image on the node since with jamming a CR-RCA run would definitely result in a verifier-side *node_compromised* inference eventually causing a news code image update with the bogus code image. In the remainder of this work we are aiming to come up with more concrete values for the above introduced parameters.

6   Trade-Off Discussion on Number of Addresses

In SWATT we face two constraining aspects for choosing a reasonable number of adresses P. The first condition is, (i) P has to be big enough such that the resulting t_Δ is larger than the network jitter. The second condition is, (ii) since the algorithm does not include the whole flash memory, there is always a chance, that malicious code segments are not attested. Consequently, P has to be chosen such that the chance for a still undetected malicious code reaches an acceptable minor percentage.

Condition 1:

$$P \cdot \left(t_{if} + t_{Reconstruction} \cdot \frac{F_{BogusCodeImage}}{F_{Total}} \right) \gg jitter$$

Condition 2:

$$1 - \left(\frac{F_{Total} - F_{BogusCodeImage}}{F_{Total}} \right)^P \gg \lambda\%$$

whereas t_{if} is the time overhead for the additional if-statement, $t_{Reconstruction}$ is the additional time overhead for the reconstruction of the original data and λ is a threshold, which should be greater than 99 %. Usually λ is represented as a multiple of σ (Gaussian distribution). If the flash memory is, in comparison to the bogus code segment, relatively big, it could be hard to fulfill the second condition. Again, P has to be increased such that a threshold λ is met. This could lead to an unpredictable large P, thus causing a large computation time per code-attestation run. On the other hand if the flash memory is relatively small, but the transport medium faces a high jitter, one could argue that $P = F$ could be the limit for P and P can not be further increased to achieve the envisioned t_Δ. However, we argue that, to allow $P > F$ for relatively small flash memory, the content of certain addresses will be hashed multiple times. Please note that, the repetition is not a security risk due to its random order. We believe that such an approach is an elegant way to allow devices with a small flash memory, to fulfill the first condition. A decently big P initially seems to be a good choice, since it increases the probability of weaving malicious code sequences of the attacker into the response hash. However increasing P also means, increasing the total amount of time which is needed to calculate the response. The advantage we gain is that each additional round also increases the overhead for the attacker. Recall that as long as the time overhead of an attacker is recognizable bigger than the jitter resulting from $t_t + t_c$, attacks can be discovered reliably at the verifier-side. Moreover, be aware that a compromised node in any case has an additional investment of two operations! The first one, the reconstruction of the data, always means considerable computational effort. However, it is only executed in a small percentage of the cases. The overall overhead of the reconstruction is compared to the overall overhead of the if-clause insignificant and therefore can be discarded. The second one is the if-clause overhead itself. The if-clause is always executed, but has a very small overhead per cycle. The cycle count P influences both operations, therefore the overhead can be calculated as follows: $P \cdot \frac{F_{BogusCodeImage}}{F_{Total}} \cdot O_{Reconstruction} + P \cdot O_{If}$, where P is the cycle count, $F_{BogusCodeImage}$ is the size of the bogus code, F_{Total} is the size of the flash drive, $O_{Reconstruction}$ is the required time to reconstruct the data and O_{If} is the overhead for the if-statement.

7 RCA-Reference Implementation and Evaluation

7.1 Tests on MSP430 with the langOS Project

To gain a deeper insight, we implemented the SWATT algorithm on an MSP430 with the IHP[1] langOS operating system. Key features of such an embedded device are, three radios, in which two are 2.4 GHz and one is below 1 GHz, an MSP430F5438A processor with 256 kB internal flash storage and 4 MB external flash storage. The MSP430 was run at 4 MHz clock rate for a low power consumption. Our exemplary code image has a size of 59 KB and 215 KB programmable memory are available on our platform. The remaining bytes are filled with pseudo random words (PRW) generated by the method we describe in Sect. 8.3. In our setup we chose a P of 2000 which translates to the probability of $4.6 * 10^{-5}$ % of not hitting an area the attacker occupies. According to λ of the Gaussian Distribution in condition 2 (Sect. 6) this is more than 4σ. One code attestation protocol run has a duration of about 1.9 s with a max. jitter of about 2000 µs. On average the code attestation hits 14 times a code piece loaded by the adversary. The additional if-statement creates an average overhead of $1 \mu s - 2 \mu s$ per cycle. Summing up the overhead, the attacker needs approximately 4000 µs more than an uncompromised node. To avoid false positives given our previous discussion, we argue that a node should not require more than $1.9 s + 2000 \mu s$. The results are shown in the Table 2. The cycle count P should be chosen, such that the overhead of an attacker is always bigger than the jitter. Thus, the only way of

---

[1] Innovations for High Performance Microelectronics.

an attacker passing the code attestation is, not bothering with code attestation and instead hoping that the RCA does not hit occupied areas. The above values do hold in a single hop scenario. Since a multi-hop scenario increases the end-to-end jitter, the parameter P has to be adapted here.

7.2   Time Based SWATT with Different Distances

With respect to the distance (see Table 4) and depending on if an obstacle is interfering the medium, the roundtrip time $2 \cdot t_t$ can differ. For example if one node is relatively close to the verifier and the other node is relatively far away located from the verifier, the first has to have a much lower threshold (calculated on the formula $2 \cdot t_t + t_c + t_\Delta$). We observe, that t_c and t_Δ, once set up are constant (see Table 3) and t_t differs for each node for its concrete deployment position. To get the t_t for each node, an initialization phase is definitely required. During this phase the verifier-node determines the t_t of the surrounding nodes. This phase should take place during the setup of the nodes. In particular, this means, if nodes migrate after the setup-phase, it could disturb the code attestation, leading to a false result. The same could happen, if obstacles are added or removed in the line-of-sight between the prover and the verifier-node. At this point we want to stress that with a purely software-based code RCA an adversary should never be such powerful to be able to reduce the distance between prover and verifier. This will always result in running bogus code undetected. Also the attacker should not be able to extend t_t during the set-up-phase. Realistically this means that the adversary should not be in physical proximity during this phase.

7.3   Effort for the If-Clause

With the help of an if-clause, an attacker needs to differentiate if the given address accesses an occupied area. For this, the adversary needs to check its lower and upper limit. The code on the MSP430 is as follows:

Table 2. Comparison between a compromised and a non-compromised node, with a test sample of more than 70 measurements.

	Valid node	Compromised node
Computation (t_c)		
min.	1800572 µs	1804891 µs
max.	1800661 µs	1804981 µs
avg.	1800636 µs	1804919 µs
roundtrip + computation$(2 \cdot t_t + t_c)$		
min	1914860 µs	1919288 µs
max	1916712 µs	1922073 µs
avg.	1915877 µs	1920281 µs
standard deviation	406 µs	411 µs

Table 3. Computation time depending on P-cycles.

computation (t_c)	1024	2048	4096	8192	16384
avg.	9.37E5 μs	1.80E6 μs	3.65E6 μs	7.91E6 μs	1.45E7 μs
standard deviation	43 μs	53 μs	59 μs	41 μs	41 μs

Note that the condition contains an upper and a lower boundary. Since both boundary conditions have to be met, it is appropriate to check whenever one condition is false. If the sequence of the conditions is chosen correctly, in the most cases just one condition has to be checked. Thus, from the adversary's perspective, if the first condition is false only the first three instructions have to be executed. These three instructions consume six CPU cycles, which translates to 1.5 μs on our reference platform. Moreover, one can observer that the closer the attacker is able to either place the malicious code at the bottom or at the top of the flash memory, the more likely only one condition has to hold. In the second case the first condition is true and the second condition is false, in which at most seven instructions have to be executed (3.5 μs). In the worst case both conditions are true and five instructions plus the reconstruction of the data have to be executed. In this case the five additional instructions do not matter, because the reconstruction of the original data is marginal more expensive.

8   Impact of the System Architecture on a CR-RCA

8.1   State Machine

It is often the case, that embedded devices do not even support pseudo parallelism, instead tasks are executed in a sequential way. A state machine is implemented, which handles interrupts and schedules tasks according to a priority system: low priority, medium priority and high priority. Idle should be the default state and each task should be associated with a priority state. After a task is finished the system always returns to the idle state. Whenever an interrupt with a higher priority state task occurs, the lower priority task is either

Table 4. Roundtrip time depending on the distance, with a test sample of 100 measurements. For an MSP430, run with 4 GHz clock rate and an 868 MHz radio.

roundtrip ($2 \cdot t_t$)	1 m	50 m	130 m
min	114470 μs	114168 μs	114574 μs
max	116882 μs	115852 μs	116568 μs
avg.	115514 μs	115168 μs	115592 μs
max. Jitter	2412 μs	1684 μs	1994 μs
root mean square deviation	450 μs	319 μs	413 μs

Algorithm 2. Additional effort in assembler instructions for the adversary.

```
CMP.W     #2,R8
JLO       ( C$L16 )
JNE       ( C$L18 )
TST.W     R9
JHS       ( C$L18 )
C$L16 :
CMP.W     #1,R8
JLO       ( C$L18 )
JNE       ( C$L17 )
CMP.W     #1,R9
JLO       ( C$L18 )
C$L17
...
...       reconstruct  data
...
C$L18
```

canceled or delayed and the new one should be started. By taking into account the paradigm of priority scheduling an CR-RCA task should always be configured with the highest priority and should ideally not be interruptible. Interrupts during the CR-RCA algorithm could lead to delaying attacks resulting in false positives.

8.2 Interrupt Handling

If an adversary is able to generate interrupts on the verifier node, he would be able to artificially increase t_c. This is true assuming that the interrupt handling compares the current task priority with the task priority of the interrupt and subsequently decides, which task to discard and which task to start/continue (See naive interrupt handling approach in Algorithm 3). Appending new tasks to a task list, could be useful from a system developer perspective, but surely increases the attack vector of an adversary. Therefore, we recommend handling incoming interrupts as simple as possible. Such an interrupt could be triggered due to incoming data frames or sensor events like temperature, blood pressure, etc. If the adversary is able to increase the verifier's t_c to a range close to the adversary's t_c, he would be able to enforce false positives. Depending on the hardware limitations, like the quantity of incoming frames per time interval the network interface can handle, this influences the whole verification process for the CA-RCA-task. Roughly speaking we can state that the more frames can be received per time interval and the lower the CPU clock rate is, the more vulnerable the embedded device will be to such type of attacks.

Algorithm 3. Naive interrupt handling

1: **if** *current_task_priority* ≥ *new_task_priority* **then**
2:     *discard new_task*
3:     *continue current_task*
4: **else**
5:     *discard current_task*
6:     *start new_task*

8.3   Pseudo Random Words

Requirements of the PRW - Motivation. Our motivation for the introduction of pseudo random words (PRW) is, to fill up the remaining free space left from the CI. To prevent compression attacks also on the PRW, the following conditions have to be ideally fulfilled:

 (i) There is no shorter representation of the PRW
(ii) The PRW should not be predictable within a reasonable effort of time

Structure of the PRW. The above two conditions are met, if (i) the PRW is not compressible and (ii) the PRW contains still too much entropy for a practical attack. To meet the condition (i) a block is defined by 256 bytes of all representations of one byte in random order. The order is defined by a secret predefined seed. Ideally the seed has to change with each software version. If more than 256 bytes are needed, multiple blocks are concatenated. This method reduces the complexity of one block from 2^{256} to 256! (see Algorithm 4). In a practical environment this should still provide an appropriate level of system security.

Algorithm 4. None Compressible Code-Algorithm

1: **procedure** CREATEDATA
2:     *rand.seed ← secret_seed*
3:     *pool ← 0...255*
4:     **while** *more Data needed* **do**
5:         *randomElement ← pool.pop(randomElement)*
6:         *data.append(randomElement)*
7:         **if** *pool is empty* **then**
8:             *pool ← 0...255*
9:     **return** *data*

9   Mandatory Recommendations for CR-RCA Implementation and Deployment

It is our objective to give mandatory implementation requirements as well as usage recommendations for a proper application of a purely software-based CR-RCA-algorithm. We emphasize that, if our recommendations are not strictly met, the CR-RCA approach will surely provide only trappy security. More concretely, it may be the case that, although the CR-RCA verifier's inference mechanism states that no bogus code is deployed on the prover's node, indeed such malicious code is yet running. However, and this is probably even more annoying, even when fulfilling all our below listed requirements, the authors of the work at hand can give no guarantee that an CR-RCA approach like e.g. SWATT provides indeed the envisioned code verification. We subdivide our recommendations into those for radio, storage, CPU and the operating system.

9.1   Recommendations for the Radio

Recommendation 1: It is mandatory to use the fastest radio during the initialization phase. In the initialization phase the t_t is set by the verifier for each prover. If an adversary is subsequently capable to use the faster radio, he could force false positives *Impact: The adversary could force false positives.*

Recommendation 2: Prover and verifier should be located in line-of-sight. Depending on the concretely chosen underlying radio technology jitter could strongly vary within transmission range. If necessary P has to be adjusted. *Impact: An obstacle within the transmission range, could highly influence the jitter whereas higher jitter leads to more false negatives.*

Recommendation 3. The jitter on the shared medium is crucial for the choice of P. Jitter heavily influences the false positive and false negative rates. Generally speaking, the higher the jitter, the larger to choose P (see Sect. 6). *Impact: The boundaries of t (including both t_c and t_t) for a valid node and a corrupted node could overlap, leading to a false positive.*

Recommendation 4: t_Δ should be chosen properly, such that the threshold does not overlap with the min. time of the adversary. A false positive is in a security sensitive scenario not acceptable, therefore the t_Δ has to be small, at the expense of the false negative rate. *Impact: If t_Δ is too small, it will not help to improve the false negative rate. On the other hand, if t_Δ is too big, it could lead to false positive.*

9.2   Recommendations for the Storage

Recommendation 5: The minimum requirements for P in a real-world environment have be ensured. In particular the conditions 1 and 2 pointed out in Sect. 6 have to be ensured.

Recommendation 6: Tertiary storage devices harmonize well with an CR-RCA approach. Tertiary storage devices usually have an access time similar to the reconstruction time; Recall that the security is based on the `if-clause` and not on the reconstruction.

Recommendation 7: The PRW should not be easily compressible or reconstructable. We thus recommend to use the proposed None Compressible Code-Algorithm (see Sect. 8.3). *Impact: Otherwise an undetected bogus code injection might be easier for an adversary.*

9.3   Recommendations for the CPU

Recommendation 8: The CPU clock rate has to be mandatorily taken into account in the calculation of P. Frequently, the CPU is underclocked to save energy. But during each CR-RCA run it is mandatory that the system runs with the maximal clock rate. That way it is guaranteed, that an attacker has the same resources available as a valid node. *Impact: Not running the CR-RCA at a maximum clock rate could lead to a false positive.*

Recommendation 9: The interrupt handling should be minimal. During the CR-RCA process the interrupt handling should be minimal. That way an adversary should not be able to influence the t_c on the verifier beyond a certain degree. (See Sect. 8.2).

9.4   Recommendations for the OS

Recommendation 10: The CR-RCA-task always requires the system's highest priority. If an adversary is enabled to interfere a running CR-RCA with interrupts or other input to be handled by the (honest) code running on the prover-side, the adversary would be able to artificially increase t_c and disguise the attack (see Sects. 8.1 and 8.2). *Impact: The adversary could use an interrupt for a higher priorized task to artificially increase the verifiers t_c and thus save time for its response computation of* ChkSum.

One may argue that the value of our recommendations is limited due to the fact that basically no configuration setting provides neither a *provable* nor at least a *pragmatic* security for CR-RCA implementation and usage. Nevertheless, we argue that the worst-case scenario is to run trappy implemented and configured CR-RCA code and at the same time believe that with such code the system security can be monitored and detected properly. If we help mitigating this with the work at hand, we feel that this is already a contribution in itself.

10   Conclusion

In the work at hand we re-visited available purely software-based remote code attestation protocols for embedded devices. Particularly, we analyzed the most prominent candidate SWATT from this class of time-bounded challenge-response protocols. Due to our prototypical implementation on the IHP sensor node based on an MSP430 microcontroller with radio interface from Texas Instruments we could derive a set of practical configuration as well as set-up recommendations which mandatorily have to be guaranteed. Otherwise SWATT (but also other protocol derivatives from this class) can definitely not ensure a proper remote code validation. On the contrary we want to stress that even with all our recommendations in place there may still be some way to bypass a purely software-based remote code attestation protocol.

Acknowledgments. The work presented in this paper was supported by the Federal Ministry of Education and Research (BMBF) within the project UNIKOPS - Universell konfigurierbare Sicherheitslösung für Cyber-Physikalische Systeme. The views and conclusions contained herein are those of the authors and should not be interpreted as necessarily representing the official policies or endorsements, either expressed or implied, of the UNIKOPS project or the BMBF.

References

1. Armknecht, F., Sadeghi, A.-R., Schulz, S., Wachsmann, C.: A security framework for the analysis, design of software attestation. In: Proceedings of the 2013 ACM SIGSAC conference on Computer & communications security, pp. 1–12. ACM (2013)
2. Castelluccia, C., Francillon, A., Perito, D., Soriente, C.: On the difficulty of software-based attestation of embedded devices. In: Proceedings of the 16th ACM conference on Computer and communications security, pp. 400–409. ACM (2009)
3. Che, W., Plusquellic, J., Bhunia, S.: A non-volatile memory based physically unclonable function without helper data. In: 2014 IEEE/ACM International Conference on Computer-Aided Design (ICCAD), pp. 148–153. IEEE (2014)
4. Elson, J., Girod, L., Estrin, D.: Fine-grained network time synchronization using reference broadcasts. ACM SIGOPS Operating Syst. Rev. **36**(SI), 147–163 (2002)
5. TCG Mobile Phone Working Group et al.: TCG mobile trusted module specification. In: Trusted Computing Group (2010)
6. Kinney, S.L.: Trusted Platform Module Basics: Using TPM in Embedded Systems. Newnes, Newton (2006)
7. Kovah, X., Kallenberg, C., Weathers, C., Herzog, A., Albin, M., Butterworth, J.: New results for timing-based attestation. In: 2012 IEEE Symposium on Security and Privacy (SP), pp. 239–253. IEEE (2012)
8. Schulz, S., Wachsmann, C., Sadeghis, A.R.: Lightweight Remote Attestation using Physical Functions, Technische Universitat Darmstadt. Darmstadt. Tech. rep., Germany, Technical report (2011)
9. Seshadri, A., Perrig, A., Van Doorn, L., Khosla, P.: Swatt: software-based attestation for embedded devices. In: Proceedings of the 2004 IEEE Symposium on Security and Privacy, pp. 272–282. IEEE (2004)
10. Stecklina, O., Kornemann, S., Grehl, F., Jung, R., Kranz, T., Leander, G., Schweer, D., Mollus, K., Westhoff, D.: Custom-fit security for efficient, pollution-resistant multicast OTA-programming with fountain codes. In: 2015 15th International Conference on Innovations for Community Services (I4CS), pp. 1–8. IEEE (2015)
11. Stecklina, O., Langendörfer, P., Vater, F., Kranz, T., Leander, G.: Intrinsic code attestation by instruction chaining for embedded devices. In: Thuraisingham, B., Wang, X.F., Yegneswaran, V. (eds.) Security and Privacy in Communication Networks. LNICSSITE, vol. 164, pp. 97–115. Springer, Heidelberg (2015)
12. Vetter, B., Westhoff, D.: Simulation study on code attestation with compressed instruction code. In: 2012 IEEE International Conference on Pervasive Computing and Communications Workshops (PERCOM Workshops), pp. 296–301. IEEE (2012)

Secure Whitelisting of Instant Messages

Günter Fahrnberger[✉]

University of Hagen, Hagen, North Rhine-Westphalia, Germany
guenter.fahrnberger@studium.fernuni-hagen.de

Abstract. Nowadays, social networks (like Facebook) have not only absorbed the function of abating pure IM (Instant Messaging) systems (like ICQ) but also their dangers, such as cyberbullying and security violations. Existent safe IM blacklist sifters already conform to the security objectives authenticity, integrity, privacy, and resilience, but simple tricks of inventive bullies can evade filtering of their originated instant messages. Therefore, this treatise introduces a novel conception of a secure IM sieve based on whitelisting. Beside a detailed view on the underlying architecture, security and performance analyses adumbrate the feasibility of the approach.

Keywords: Authenticity · Blind computing · Censorship · Cloud · Cloud computing · Cyberbullying · Cyberharassment · Cyberstalking · Filter · Flaming · Instant Messaging · Instant messenger · Integrity · Privacy · Protection · Resilience · Safety · Screener · Secret sharing · Secure computing · Security · Sieve · Sifter

1 Introduction

These days, Internet users can swap information in subsecond timescale. IM (Instant Messaging) represents the most prominent way to transact communication between two or more people. Luckily, on the one hand, chatters in democracies can freely express their opinions in instant messages without needing to fear political persecution and punishment. Unfortunately, on the other hand, evildoers exploit this freedom to bully their victims through instant messaging. Cyberbullying denotes deliberate repeated harm or harassment in online communication media. It occurs in variously severe sorts. Already undesired online contacting (via instant messengers) can be considered as cyberstalking. Flaming, the use of assaultive or rude language in heated or intense instant messages, implies hostile and insulting interaction among humans. Cyberharassment means offensive instant messages targeted at one or several individuals. Cyberstalking, flaming, and cyberharassment exemplify just three kinds of cyberbullying. Blocking the causing IM accounts seems to be the most obvious approach to inhibit further affronts, but nothing can prevent the miscreants to spawn new accounts, (re)inspire the mobbed sufferers with trust, and continue to afflict them. A more promising resort would be a text filter that inspects all instant messages addressed to cyberbullying-endangered IM receivers. The sifter either

© Springer International Publishing AG 2016
G. Fahrnberger et al. (Eds.): I4CS 2016, CCIS 648, pp. 90–111, 2016.
DOI: 10.1007/978-3-319-49466-1_7

cleans objectionable messages or drops them before their consignees get hold of them. Of course, adult addressees, respectively legal guardians of minor recipients, must consciously agree with such a screening rather than being forced to. If they have decided to employ a text sieve, then it shall always be on duty irrespective of their location and used IM client application. Aside from location-independence, they anticipate a sieve that also includes colloquialisms, vogue terms, and cutting-edge swearwords in its decisions. These two requirements motivate a centralized, updatable screening solution (best directly coupled with the IM core platform) with proper resilience rather than a trivial client-side approach.

At this point, the first privacy problem arises. Public clouds usually shelter IM core platforms due to their demand of scalability for the incalculable user growth. Thus, affiliated screeners would also be located in public clouds. Cloud providers, intelligence agencies, and even the IM operators themselves have a field day retrieving private data out of instant messages for individual and big data analyses, to modify, or to stash them away. If IM users knew the risk for their confidential data in cloud services, then they would never opt for (online filtering of their) instant messages. Hence, a prudent sifting design (abetting not only privacy but also authenticity, integrity, and resilience) must do the trick.

Related work in Sect. 2 shows feasible secure blacklisting of instant messages. Blacklisting connotes the search and eradication of unsolicited terms. Unfortunately, already explicit phrases in spaced form can circumvent any blacklist filters, not to mention resourceful fake character insertions and uppercase-lowercase-combinations [32].

For this purpose, this paper comes along with a secure whitelisting technique that scans instant messages securely and forwards only those ones with solely harmless words to their destinations. The novel creation amalgamates the following principles:

- **White- instead of Blacklisting to impede circumvention:** Only instant messages with merely approved words inclusive their declined or conjugated variants become delivered to their intended targets.
- **Secret sharing between IM relay and IM filter to maintain privacy:** An IM relay becomes only aware of the source and the destination accounts of instant messages, but not of their plaintext content. An IM filter just processes a huddle of addressless character strings encrypted symmetrically with the secret key of the IM relay.
- **PKI (Public Key Infrastructure) to sustain anonymity and authenticity:** A consigner device (hereinafter called Alice) encrypts an accrued whole plaintext instant message and its bulk of individual words separately in a hybrid ilk, i.e. it enciphers the instant message symmetrically with a random secret key and the heap of its individual words with another secret key before it veils the secret keys with the public keys of the addressees. Hybrid cryptography joins the velocity and unlimited input length of symmetric ciphering and the benefit of unequal keys for encryption and decryption through asymmetric cryptographic functions. Moreover, Alice convinces the IM consignee

terminal (thereinafter named Bob) of its authenticity by signing a hash value of the ciphertext instant message with its private key. Altogether, Alice remains anonymous for the sieve and can be sure at the same time that merely the possessors of the appropriate private keys can confidently decrypt the ciphertexts.

- **Training mode to enforce resilience:** Bob can recommend vocables covertly to the IM filter which should be included in the whitelist in future or excluded from there.
- **One-time secret keys to assure privacy:** Alice and Bob draw on a unique arbitrary secret key for the symmetric encipherment of each instant message respectively on another one for the symmetric scrambling of a batch of individual words to reach nondeterministic encryption and to avert collision attacks which would be a consequence of ciphertext repetitions.
- **Onionskin paradigm to warrant authenticity, integrity, and privacy:** A topical hybrid cryptosystem between all involved components ensures an extra umbrella against menaces during data transport.

In respect to existing literature, Sect. 2 reviews technical contemporary instruments and interdisciplinary nontechnical scientific stances referring regulation of online safety especially for children and adolescents. Section 3 proposes an innovative technical safeguard that satisfies the needs of stakeholders for IM security. As Sect. 4 proves the invulnerability of the tool by dint of a stalwart security analysis, Sect. 5 compares its performance with that a proposal mentioned in Sect. 2. The upshot of this disquisition becomes summarized and discussed in Sect. 6.

2   Related Work

Habitually, technical treatises purely cite other technical papers and articles. This one rather chooses an interdisciplinary direction by bearing on nontechnical viewpoints in Subsect. 2.1 and technical merits in Subsect. 2.2.

2.1   Nontechnical Related Work

The majority of scholars from a variety of research areas, but with fair literacy and a common interest in online protection for underage people, shares the position that society ought to vest user groups at risk with adequate skills to cope online hazards rather than to enact restrictive provisions [2]. In compliance with this prevailing attitude, the suggested whitelist checker poses a reasonable implement for voluntary exertion by responsible-minded youngsters or parents rather than by a lawful mandatory measure. It goes without saying that such an idea can just become a successful story with the will of the leading IM offerers.

Thierer advocated *resiliency* and *adaption* as optimal response strategies to counter online safety risks [29]. Specifically, education equips guardians with the media to guide the mentoring process and their kids with a better preparation for inevitable surprises rather than *prohibition* or *anticipatory regulation*. The recommended whitelist filter constitutes such a supportive medium.

Van den Berg criticized techno-regulatory appliances (inter alia parental control, filtering content, browsers for kids, using ports and zoning) as hindrances for teenagers' ability to explore and experiment with the (positive and negative) potential of the Internet freely [1]. She called for a move away from them towards persuasive technologies and nudging solutions as a worthwhile contribution to improve child safety on the Internet. Persuasion of a higher risk-awareness does not need to conflict with techno-regularization, e.g. simultaneously knowing the facts of cyberthreats and exercising techno-regulative practices.

Sonck and de Haan took the same line by seeing the amelioration of digital skills as a form of empowerment and a possible endeavor to online safeness for youths [27]. Despite that, empirical studies had not found that capabilities mastering challenges in the Internet significantly cause the impediment of unwanted experiences.

Notten substantiated the impact of families and societies on juveniles' risky online behavior [25]. She propounded higher parents' encouragement to confine their offsprings' Internet use, in particular in modern countries where parenting styles appear often more permissive. The advised whitelist screener complies with Notten's request.

Due to the iron memory of the Internet, van der Hof as well as Ciavarella and De Terwangne commemorated the important right to oblivion, also referred to as the right to be forgotten [5,18]. It means that in different circumstances data owners gain the possibility to obtain the deletion of their personal data (such as instant messages and IM accounts), in sundry data protection directives, mainly for young people who want to restart with a clean slate. The commended whitelist sieve innately obeys these laws because saved ciphertext messages in clouds secured by a strong cryptosystem should not be revealable for decades.

Amongst others, Lievens characterized four empowerment strategies to help reducing peer-to-peer risks in social networks: enhancing media literacy, proffering information, establishing efficient reporting mechanisms, and peer-to-peer strategies [22]. The introduced whitelist system facilitates ad hoc reports about passed and blocked instant messages for curious custodians.

Van der Zwaan et al. comparatively assessed the effectiveness of the successional present Internet safety precautions against cyberbullying by means of diverse criteria: content and behavior analysis, filtering, monitoring, blocking of undesirable contacts, content reporting, age and identity verification, and educational technology [32]. In spite of their preference of blocking unwanted contacts, they claimed an amalgamation of social, legal, and technological measures to achieve optimum results. Their classification of (content) sieving as a compulsive measure may be objectionable if it voluntarily happens.

Genner pleaded for protecting adolescents by fostering their resilience with the aid of education rather than through legislation [17].

Beside the periled age bracket, Vandebosch's multi-stakeholder model urges school staff, caregivers, the police, ISPs (Internet Service Providers), news media, policymakers, and researchers to adopt proactive roles against cyberbullying [30].

Van der Knapp and Cuijpers demonstrated that merely a small proportion of children becomes victimized by online sexual solicitation during their lifetime [20]. However, they asked for future research regarding the interrelationship between grooming and psychosocial development to attain insights into risk and protective factors.

2.2 Technical Related Work

At the outset of this subsection, it needs to be said that the current literature contains less technical contributions about pure instant message purgers than one might expect. One reason may be the favoritism of sensitization for jeopardies in the Internet over regulations, another one the tendency to comprehensive safeguarding. Anyway, this subsection chronologically describes some dedicated instant message cleaners, followed by their shortcomings that make a seminal resolution much-needed.

In 2004, Boss and McConnell have patented a system and method for filtering instant messages [3]. An originator sees a list of the recipient's contexts and must match the outgoing message to one of them. This suggestion cannot detain bullies to select nonrestrictive contexts at their own choice in order to resume their harassments. That is why it induced ineffectiveness against cyberbullying.

One year later, in 2005, Liu et al. pitted themselves against spim (IM spam) with a hierarchical detection and filtering architecture [23]. It defends receivers against unsolicited junk messages and DoS (Denial of Service) assaults. Their concept focuses on the suppression of automatically generated messages with multiple filters and, thence, cannot guard bullied folk.

Landsman's patent from 2007 pursued the thoughts of Liu et al. and additionally remarked the spimmer's ease of becoming a trusted message source, e.g. by stealing the IM account credentials of respectable users and impersonating them [21]. On these grounds, Landsman concentrated on an obligatory verification process for each sender to hamper automated spim. Nonetheless, his remedy cannot inhibit cyberbullying.

Ganz and Borst discerned the porousness of blacklisting in 2012 and came out with a patent-registered invention that merges white- with blacklisting [16]. Their contrivance indicates a good trend, but it purely screens plaintext messages and, for this reason, allows IM hosts to take advantage of them easily.

In 2013, Fahrnberger overcame this deficit with his homomorphic cryptosystem SecureString 1.0 that perpetuates the secrecy of encrypted character strings even during modifications on them [7]. Already in the same year, he issued his advanced successor SecureString 2.0 which resolved emerged functional and security-relevant defects [8]. A continuing publication underpinned the grasp of SecureString 2.0 with security and performance analyses [9]. SafeChat was published as a blacklisting chat filter relying on SecureString 2.0 [14]. The timeline went on with SIMS (Secure Instant Messaging Sifter) as an improved blacklist-based realization of SecureString 2.0 [10]. In 2015, Fahrnberger himself disqualified SecureString 2.0 due to its discovered weakness against repetition pattern offenses [11]. This finding led to the evolution of SecureString 3.0 that avoids any ciphertext repetitions and, in addition, scores by bearing up authenticity, integrity, and resilience as well [13]. A book chapter about implementation details, security and performance issues supplemented the knowledge of Secure-String 3.0 [12].

Even a draft for blacklisting based on SecureString 3.0 cannot preclude that savvy culprits slightly alter blacklisted expressions just to keep them understandable for readers and to get them through to their victims' eyes. For that reason, a credo change towards whitelisting becomes inescapable.

The next section expatiates on the architecture of a plan for a homomorphic IM scanner depending on whitelisting.

3   Architecture

This section sheds light on constructional subtleties with the help of two subsections. While Subsect. 3.1 attends to a description for the basic components of the proposed IM filter system, Subsect. 3.2 specifies the flows of instant messages and for initialization.

3.1   Components

Figure 1 depicts a rudimentary graphical overview of the necessary constituents for the secure whitelisting of instant messages. Attentive readers immediately perceive the similarity of the architectural picture with that one of SIMS [10]. Notwithstanding the resemblance to SIMS, the whitelist construction names the IM core platform as IM relay, and the IM filter replaces the TTPG (Trusted Third Party Generator). In contrary to the separate high-speed WAN (Wide Area Network) between the IM core platform and the TTPG in SIMS, the whitelist method does not inclose such a special bus between the IM relay and the IM filter, because it manages without the time- and bandwidth-consuming generation and dissemination of a filter repository for each treated instant message. Furthermore, the achievement of nondeterministic encipherment necessitates an on-board RRNG (Real Random Number Generator) preferably, or PRNG (Pseudo Random Number Generator), at least seeded with real random data, in every IM client rather than merely in the IM relay.

Transmission. Transmission allegorizes the backbone of IM topology as a cloud of noticeably opaque pathways to interconnect all the labeled items in Fig. 1. This obscurity in coincidence with lack of control impairs the trustworthiness of the transmission cloud. Due to the bad need of insecure communication paths besides secure ones for global reachability of IM clients, applied cryptography must obliterate this drawback.

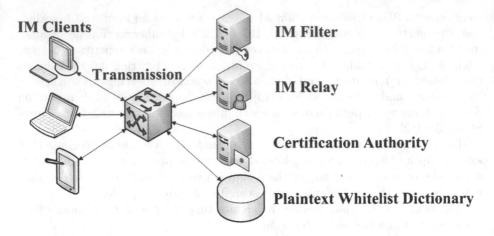

Fig. 1. Components for secure whitelisting of instant messages

IM Clients. The three illustrated IM clients in Fig. 1 act as representatives for a multitude of miscellaneous devices in place all over the world with installed IM software and online functionality. Logically, all clients operate bidirectionally, i.e. they incorporate applicability as IM senders as well as IM receivers. Accordingly, they have to encrypt leaving and to decipher arrived instant messages. As aforementioned, safe emitting of instant messages demands the IM clients to produce arbitrary data for nondeterministic ciphering. Receiving IM clients support two mutually exclusive operational modes: real and training mode. While in real mode addressees do not become aware of instant messages that were dropped in the IM relay, the training mode permits them to even see the dubious ones in order to propound terms to the IM filter that should be included to the whitelist in future or excluded from there.

Certification Authority. The CA (Certification Authority) inspires confidence in the identity of truthful sending IM clients to their addressed receiving IM clients by building the heart of a common PKI (Public Key Infrastructure). Thereby, it also makes IM recipients aware of spurious instant messages. Its assignment comprises issuing certificates for IM clients by signing their certificate requests, revoking compromised certificates, validating authentic and falsifying void ones.

Plaintext Whitelist Dictionary. The plaintext whitelist dictionary appears as a steadily available online database with the unencrypted unobjectionable character collections of all supported languages. It must also embrace the digits from zero to nine to let the IM filter permit instant messages with numeric parts, like times, dates, and other numbers. Supplementarily, the database supplier must promptly update the whitelist vocabulary with newly emerging, innocuous expressions to shirk upset writers whose texts become wrongly repudiated.

Surely, it needs convenient defense from DoS strikes and unauthorized manipulations to serve as a reliable spring for dependent IM filters.

IM Relay. The semi-honest(-but-curious) [4,26] IM relay primarily takes the responsibility to store and remit incoming instant messages to their designated aims. In the presented constellation, it forms a heterogeneous secret-sharing cluster together with the also semi-honest(-but-curious) IM filter. Apart from the exchanged data within the defined flows in Subsect. 3.2, each of both cluster entities must run in its own isolated environment to keep up privacy. In the current setting, the IM relay confronts the associated IM filter with the pack of delimited ciphertext words for each received instant message to let the IM filter scrutinize their appearance in the ciphertext whitelist. Beyond that, the IM relay periodically interrogates the plaintext whitelist dictionary for changes. If applicable, it fetches an updated copy of the whitelist in order to symmetrically encipher all dictionary items individually with its secret key and to convey the ciphertexts to the IM filter for further processing.

IM Filter. The IM filter as counterpart of the IM relay in their formed secret-sharing cluster can be denominated as a simple ciphertext comparator and may be placed in a tamper proof microprocessor [31] of the IM relay.

Generally, its principal task consists of announcing the containedness of ciphered character strings (gotten from the IM relay) in the also-enciphered whitelist to the IM relay. More precisely, the IM filter solely throws positive acknowledgments for instant messages that exclusively comprise of whitelist words. Each time before the IM sifter can execute such comparisons, it must release the hybrid encryption of a processed word packet with its private key, which an Alice has imposed with the fitting public key.

As auxiliary job, the IM filter receives suggestions of IM clients in training mode with expressions that ought to be added to its enciphered whitelist or deleted from there. It adds a new buzzword to its (ciphertext) whitelist if the number of trustworthy advocates exceeds a predefined threshold. In the same way, it removes a whitelist entry in the event of sufficient opposers. Trustable are those proponents and objectors whose rate of barred outbound instant messages does not rise above a predetermined percentage.

3.2 Flows

This subsection bends to the conducted communication between the previously explained units.

The initialization flow intentionally performs once prior to any transfers of instant messages. It solely reruns for secret key updates of the IM relay.

By contrast, the instant message flow recurs for every transmitted instant message. Over and above all alluded finesses in this subsection, both flows engage up-to-date hybrid transport cryptography to shield the entire communication

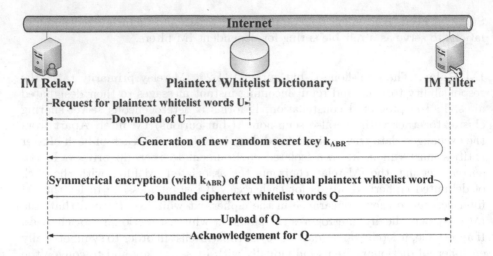

Fig. 2. Initialization flow for secure whitelisting of instant messages

from security breaches. Otherwise, for example, a villain could secretly manipulate plain- and ciphertext whitelist words during haulage or counterfeit IM addressers by mounting replay raids.

Initialization Flow. The IM relay triggers the initialization flow shown in Fig. 2 by querying the plaintext whitelist dictionary for an actual snapshot U. Upon receipt, the IM relay randomly distills a fresh secret key k_{ABR} and applies it by symmetrically enciphering every plaintext whitelist word $u_{\#b} \in U|1 \leq b \leq |U|$ to $E(u_{\#b}, k_{ABR}) = q_{\#b}$ individually. Because of randomized padding, the encryption scheme scrambles all whitelist character strings to ciphertexts with single or multiple block size. Eventually, the IM relay lexicographically arranges them to obfuscate their original order and supplies the IM filter with the reordered bunch $Q = \{q_{\#b}|1 \leq b \leq |U|\}$.

Instant Message Flow. Every IM client sparks off the sketched flow in Fig. 3 and turns into an Alice when it has to send an instant message to an IM addressee Bob.

First of all, Alice retrieves the currently valid secret key k_{ABR} from the IM relay and supplementarily creates two arbitrary secret keys k_{AB} and k_{ABF}. Then Alice splits the concerned plaintext instant message v into its unique words $v_{\#1}, \cdots, v_{\#b}, \cdots, v_{\#c}$ (viz. she eradicates occurring duplicates) on the basis of their delimiters (exempli gratia blanks or tabulators) and separately encrypts each $v_{\#b}|1 \leq b \leq c$ with k_{ABR} and a fashionable symmetric cryptosystem to $w_{\#1}, \cdots, w_{\#b}, \cdots, w_{\#c}$. Subsequent lexicographical sorting of the ciphertexts $w_{\#1}, \cdots, w_{\#b}, \cdots, w_{\#c}$ to $\{w_{\#b}|1 \leq b \leq c\}$ hides their original order. On top of that, Alice cryptographically secures the bundle of ciphertext words $\{w_{\#b}|1 \leq b \leq c\}$ once more with k_{ABF} and a suitable symmetric encryption scheme to

$w' = E(\{w_{\#b}|1 \le b \le c\}, k_{ABF})$. While the first scrambling with k_{ABR} disguises the character strings for the IM filter, the second one with k_{ABF} conceals them for the IM relay. This legitimates the double encryption effort. The full instant message v itself needs to be purely conveyed to Bob rather than to be queried or edited. On that account, Alice symmetrically enciphers v with k_{AB} to $w = E(v, k_{AB})$. Thereafter, Alice prepares k_{AB} and k_{ABF} to such an extent that exclusively the IM sieve can utilize k_{ABF}, and Bob can apply both k_{AB} and k_{ABF}. On account of this, Alice asymmetrically encrypts k_{AB} respectively k_{ABF} with Bob's public key B_{pub} to $X(k_{AB}, B_{pub})$ respectively $X(k_{ABF}, B_{pub})$, and again k_{ABF} with the public key of the IM sifter F_{pub} to $X(k_{ABF}, F_{pub})$. Additively, Alice attests the genuineness of w by including a corresponding digital fingerprint $H(w)' = X(H(w), A_{priv})$ which enfolds the hash value of the ciphered instant message $H(w)$ signed with her private key A_{priv}. Alice concludes her mission by conveying $w, w', X(k_{AB}, B_{pub}), X(k_{ABF}, F_{pub}), X(k_{ABF}, B_{pub}), H(w)'$, and A_{cert} to the IM relay.

The IM relay refers w' and $X(k_{ABF}, F_{pub})$ to the IM filter. The IM filter restores $k_{ABF} = X^{-1}(X(k_{ABF}, F_{pub}), F_{priv})$, i.e. by deciphering $X(k_{ABF}, F_{pub})$ with its private key F_{priv} and releases the outer protective layer of w' with k_{ABF} to $\{w_{\#b}|1 \le b \le c\} = D(w', k_{ABF})$. Afterwards, the IM filter seeks each yet symmetrically ciphered word $w_{\#b}$ in its ciphertext whitelist dictionary Q and notifies the IM relay whether all pieces of an instant message implicate a hit or not. Regardless of the notification output, the IM relay fires an acknowledgment back to Alice.

Solely in the event of a 100% whitelist hit rate in real mode, the IM relay routes $w, w', X(k_{AB}, B_{pub}), X(k_{ABF}, B_{pub}), k_{ABR}, H(w)'$, and A_{cert} to Bob accordant Fig. 4. Else, the IM relay terminates the suspicious instant message.

Bob first needs to verify the originality of w. On this account, Bob lets the CA confirm/reject Alice's certificate A_{cert}. In case of a positive answer, Bob deciphers the digital fingerprint $H(w)'$ with Alice's public key A_{pub} (that is part of Alice's certificate A_{cert}) and opposes the outcome $X^{-1}(H(w)', A_{pub})$ to the self-deduced hash value $H(w)$ of w. Even if $X^{-1}(H(w)', A_{pub})$ and $H(w)$ accord, Bob still does not have the guarantee of a gotten inoffensive instant message because Alice could have embezzled words of v in w'. On account of that, Bob double-checks whether the IM screener obtained all individual character strings of v in w' for audit. Thereto, Bob reconstructs $k_{AB} = X^{-1}(X(k_{AB}, B_{pub}), B_{priv})$ and $k_{ABF} = X^{-1}(X(k_{ABF}, B_{pub}), B_{priv})$ through asymmetric decipherment with his private key B_{priv}. Utilizing k_{AB} for the symmetrical decryption of w leads to $v = D(w, k_{AB})$. In the same manner as Alice did, Bob fragments v into its constituent words and erases all appearing duplicates in oder to come by $\{v_{\#b}|1 \le b \le c\}$. Subsequently, Bob derives $\{v'_{\#b}|1 \le b \le c\}$ by double symmetric decryption $D(D(w', k_{ABF}), k_{ABR})$. Only if $\{v'_{\#b}|1 \le b \le c\}$ embodies all elements of $\{v_{\#b}|1 \le b \le c\}$, Bob acquires the guaranty that the IM sieve vetted the total instant message and displays v.

Finally at any rate, Bob pushes a notice of receipt to the IM relay.

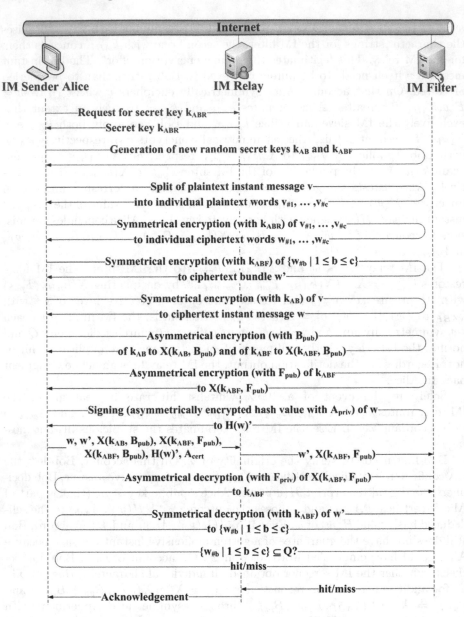

Fig. 3. Instant message flow for secure whitelisting

Once Bob has informed the IM relay of being in training mode, the IM relay also delivers every suspect ciphertext instant message w to Bob conjointly with w', $X(k_{AB}, B_{pub})$, $X(k_{ABF}, B_{pub})$, k_{ABR}, $H(w)'$, and A_{cert}. Bob executes the same steps as in real mode according to Fig. 4, but it proceeds with the additional ones in Fig. 5 in lieu of a frugal acknowledgment backwards to the

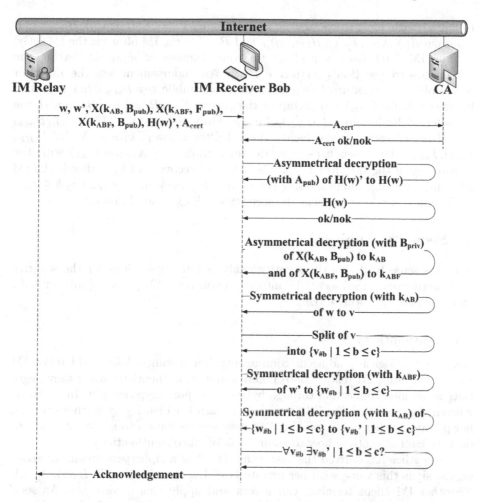

Fig. 4. Instant message flow after secure whitelisting in real mode

IM relay. Firstly, the human user behind Bob can suggest a set of expressions $\{v_{W\#b}|1 \leq b \leq c_W\}$ to be whitelisted in future and another set $\{v_{B\#b}|1 \leq b \leq c_B\}$ not to be whitelisted anymore. The successive action entails a fresh secret key k_{BF}. After this, Bob symmetrically scrambles expression by expression with k_{ABR} and creates $\{w_{W\#b}|1 \leq b \leq c_W\}$ out of $\{v_{W\#b}|1 \leq b \leq c_W\}$ and $\{w_{B\#b}|1 \leq b \leq c_B\}$ out of $\{v_{B\#b}|1 \leq b \leq c_B\}$. After that, Bob safely packages both ciphertext sets with symmetrical encipherment by the use of k_{BF} to $w_{WB} = E(\{\{w_{W\#b}|1 \leq b \leq c_W\}, \{w_{B\#b}|1 \leq b \leq c_B\}\}, k_{BF})$. Beyond Bob, merely the IM filter may restore $\{w_{W\#b}|1 \leq b \leq c_W\}$ and $\{w_{B\#b}|1 \leq b \leq c_B\}$ out of w_{WB}. As a consequence, Bob wraps k_{BF} up in $X(k_{BF}, F_{pub})$ through asymmetrical ciphering with the public key of the IM filter F_{pub}. Ultimately, Bob adduces evidence for the realness of w_{WB} by signing the hash value

$H(w_{WB})$ with his private key B_{priv} to $H(w_{WB})' = X(H(w_{WB}), B_{priv})$ and sends $w_{WB}, X(k_{BF}, F_{pub}), H(w_{WB})'$, and B_{cert} to the IM filter via the IM relay.

The IM filter assures itself of the veritableness of w_{WB} by letting the CA endorse/refuse Bob's certificate B_{cert}. An endorsement lets the IM filter descramble the signature $H(w_{WB})'$ with Bob's public key B_{pub} (that is part of Bob's certificate B_{cert}) and juxtapose the output $X^{-1}(H(w_{WB})', B_{pub})$ with the self-inferred hash value $H(w_{WB})$ of w_{WB}. Only if both $X^{-1}(H(w_{WB})', B_{pub})$ and $H(w_{WB})$ completely coincide, the IM filter recovers $k_{BF} = X^{-1}(X(k_{BF}, F_{pub}), F_{priv})$, i.e. through asymmetric descrambling of $X(k_{BF}, F_{pub})$ with the private key of the IM filter F_{priv}. The achieved recovery of k_{BF} affords the IM filter unscrambling and processing $\{w_{W\#b}|1 \le b \le c_W\}$ and $\{w_{B\#b}|1 \le b \le c_B\}$.

The next section evidences the security of all explicated flows.

4   Security Analysis

It makes sense to divide the security analysis into subsections for the security goals authenticity (Subsect. 4.1), integrity (Subsect. 4.2), privacy (Subsect. 4.3), and resilience (Subsect. 4.4).

4.1   Authenticity

Each pair of logically adjacent components (for instance Alice – IM relay, IM relay – IM filter, IM relay – Bob) inherits mutual authenticity when they negotiate a common connection through hybrid transport cryptography. In contrast, Alice makes sure that the IM filter cannot learn her identity by neither exerting her private nor her public key for the safety encapsulation to w'. What is left to do is to circumstantiate how to assure Bob of Alice's authenticity.

After Alice has derived the hash value $H(w)$ of a ciphertext instant message w, she signs this value with her private key A_{priv} to $H(w)' = X(H(w), A_{priv})$. No other IM client terminal can access and apply this private key. As soon as Bob gathers the signature $H(w)'$, he lets the CA affirm Alice's certificate A_{cert} validity and unleashes the hash value $H(w)$ with Alice's public key A_{pub} enclosed in her certificate. In the last step, Bob computes the hash value $H(w)$ of the gained ciphertext instant message w as well in order to check its equality with the unleashed one. If both hash values resemble, then Bob can feel confident of Alice's authenticity. The computation of the hash value of w in lieu of that of v shall designedly not comply to the Horton principle *Authenticate what is being meant, not what is being said!* [15] because a distinct secret key k_{AB} for the scrambling of v to w dodges recurring signatures even in the case of repetitive plaintext instant messages.

The identical mechanism comes into operation to persuade the IM filter of Bob's authenticity if Bob suggests any terms in training mode.

4.2   Integrity

End-to-end block encryption with the arbitrary secret key k_{AB} and Bob's public key B_{pub} protects the integrity of an instant message on its complete path.

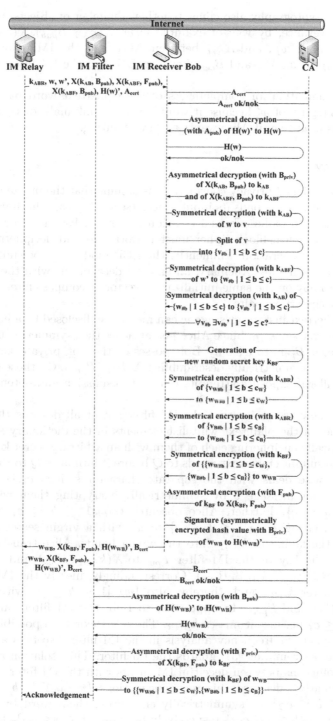

Fig. 5. Instant message flow after secure whitelisting in training mode

Such hybrid cryptography also clings together a number of objects during their common conveyance, by way of example $w, w', X(k_{AB}, B_{pub}), X(k_{ABF}, F_{pub}),$ $X(k_{ABF}, B_{pub}), H(w)',$ and A_{cert} between Alice and the IM relay, or $w_{WB},$ $X(k_{BF}, F_{pub}), H(w_{WB})',$ and B_{cert} between Bob and the IM filter via the IM relay.

Bob's examination for completeness of the ciphertext word parcel w' preserves the integrity of w' because it would even debunk malicious omittance of parcel words in w', e.g. if Alice wants to thwart sifting.

4.3   Privacy

The below-mentioned privacy considerations assume that the utilized asymmetrical encryption scheme pads its input blocks (secret keys and hash values) with haphazard data to baffle effectual collision offensives. By the same token, the exercised symmetrical block cipher must perform en- and decipherments with any safe mode of operation. Eminently, the ECB mode [19] does not belong to the group of safe modes. Ancillary, all en- and decryptions with the secret key k_{ABR} have to rely on the same initialization vector to compass filterability with fix plaintext-ciphertext-pairs.

Each ciphertext instant message w can merely be disclosed by a holder of the appendant secret key k_{AB}. Since Alice just attaches the asymmetrically ciphered version $X(k_{AB}, B_{pub})$ to w, only Bob possesses the apt private key B_{priv} to regain k_{AB} by asymmetrically descrambling $X(k_{AB}, B_{pub})$. On those grounds, w endures all illegal exposure attempts during transmission and detention in the IM relay.

The IM relay ascertains that the IM filter not at all descries the meaning and the order of the plaintext whitelist elements in the dictionary snapshot U by symmetrically enciphering each of them with an arbitrary secret key k_{AB} and sorting the resultant ciphertext words to Q before it intimates Q to the IM filter.

Alice likewise deals with the separate character strings $v_1, \cdots, v_{\#c}$ of a plaintext instant message v by symmetrically encrypting them with k_{AB} to $\{w_{\#b}|1 \leq b \leq c\}$. Further, Alice obscures $\{w_{\#b}|1 \leq b \leq c\}$ for the IM relay by converting $\{w_{\#b}|1 \leq b \leq c\}$ to w' with a virgin secret key k_{ABF}. To obviate the falling of k_{ABF} into the wrong hands, Alice transforms k_{ABF} with the public key of the IM filter F_{pub} to $X(k_{ABF}, F_{pub})$ and with Bob's public key B_{pub} to $X(k_{ABF}, B_{pub})$. Correspondingly, merely the IM filter and Bob can recover k_{ABF} and, thereby, also $\{w_{\#b}|1 \leq b \leq c\}$ with their private keys F_{priv} and B_{priv}. In opposition to Bob, the IM filter cannot unveil $\{v_{\#b}|1 \leq b \leq c\}$ because it misses k_{ABR}. The sole theoretical possibility of conducting a ciphertext frequency analysis in the IM filter also fails as long as a tamper proof surrounding [31] guards the IM filter. This isolation additionally forecloses collusions between deceitful IM clients with the IM filter.

Bob similarly safeguards his recommendations $\{v_{W\#b}|1 \leq b \leq c_W\}$ and $\{v_{B\#b}| 1 \leq b \leq c_B\}$ by symmetrically encrypting them word by word with k_{ABR} to $\{w_{W\#b}|1 \leq b \leq c_W\}$ respectively to $\{w_{B\#b}|1 \leq b \leq c_B\}$. He symmetrically enciphers the conglomerate of both sets with an unexploited secret key

k_{BF} to w_{WB} and commutes k_{BF} into $X(k_{BF}, F_{pub})$ (with the public key of the IM filter F_{pub}). The consequence is that solely the IM filter can recapture k_{BF} and, therewith, the ciphertext sets $\{w_{W\#b}|1 \leq b \leq c_W\}$ and $\{w_{B\#b}|1 \leq b \leq c_B\}$, but not the plaintext sets $\{v_{W\#b}|1 \leq b \leq c_W\}$ and $\{v_{B\#b}|1 \leq b \leq c_B\}$.

4.4 Resilience

All core systems (CA, IM relay, IM filter, and plaintext whitelist dictionary) work load-independently as long as the selected cloud provides scalability concerning tamper proof microprocessors. Advisably, their resilience can be increased through site mirroring.

Right-minded transmission operators also pay attention to provide their customers with failsafe routing via an amount of redundant links.

Solely, the selective replication of specific IM client devices would be preposterous because all users with broken terminals can just as well return to their IM accounts with any functioning replacement.

The next section complements the security analysis of the advocated whitelist sifter with a proof of its viability through a prudential performance analysis.

5 Performance Analysis

Due to the deficiency of the described safe whitelist searcher in live operation, measurements during a correspondent experiment had to give conformable performance insights. While Subsect. 5.1 outlines the deployed experimental buildup, Subsect. 5.2 exposes the yield of the experiment.

5.1 Experimental Setup

The employed experimental setup resembles that one of SecureString 2.0 [9], viz. an IM relay communicates with Alice, Bob, and an IM filter via Java RMI (Remote Method Invocation) in accordance with the needed flows for real mode in the Figs. 2, 3, and 4. Bob analogously invokes a CA via RMI for the investigation of Alice's submitted certificate A_{cert} (see Fig. 4). Each of the above-mentioned six parties (plus a Java remote object registry) became simulated through a separate Java 8 process on an HP DL380 G5 server with eight 2.5 GHz cores and 32 GB main memory that operates Fedora Core 64 bit Linux. The simulation deliberately happened on a lonesome computer because it did not intend the measurement of volatile network latencies. Each Linux process for the trial ran with the lowest nice level and, therefore, with the highest priority to eliminate perturbing effects caused by concurrently launched processes. Alice, Bob, the IM relay, and the IM filter availed themselves of AES-128/CBC/PKCS5Padding [6,19] as symmetrical cryptosystem and ECC (Elliptic Curve Cryptography) [28] with the three differing key sizes 192, 224, and 256 bits as asymmetrical one.

Which role has each unit to take on during the test phase?

Plaintext Whitelist Dictionary. The plaintext whitelist dictionary bases upon a MariaDB with a primary table of 128,804 genuine English vocables from WordNet [24] and a secondary one with a proper subset of 128,656 nonhazardous elements. As a result, the larger table as well encloses 148 questionable words mostly applied by bullies to annoy their victims.

Java Remote Object Registry. The execution of the shell command *rmiregistry* sparks an instance of the Java remote object registry that accepts registrations and inquiries of Alice, of Bob, of the CA, of the IM relay, and of the IM filter.

Certification Authority. The CA inscribes itself in the Java remote object registry in the beginning and stands by for Bob's queries.

IM Receiver Bob. Upon instantiation, Bob enrolls as an ordinary subscriber in the Java remote object registry and waits for feed from the IM relay in order to exert the real mode flow in Fig. 4.

IM Filter. The IM filter instance enlists in the Java remote object registry after its start and offers the IM relay procedures to assimilate and browse ciphertext whitelists.

IM Relay. The created IM relay signs on in the Java remote object registry and initializes the IM filter and itself pursuant to the initialization flow in Fig. 2.

IM Sender Alice. Alice takes off as the last component, just as with the inscription in the Java remote object registry. Afterward, she evaluates the mean cycle time of single-word-containing instant messages by iterating the submission flow in Fig. 3 1,000,000 times with arbitrarily picked vocables from the primary table of the plaintext whitelist dictionary U. Alice repeats this measuring for instant messages comprising $2 \leq c \leq 7$ delimited words with a sample size of 1,000,000 per word count as well. The probability of having at least one obnoxious item in an instant message with c words equals

$$P(c) = 1 - \prod_{b=1}^{c} 1 - \frac{|\{\text{bad words in } U\}|}{|U|}$$

$$= 1 - \left(1 - \frac{|\{\text{bad words in } U\}|}{|U|}\right)^{c}$$

$$= 1 - \left(1 - \frac{148}{128,804}\right)^{c} \approx 1 - 0.9989^{c}$$

Table 1 itemizes the resultant average percentage of illicit instant messages that must be spotted by the IM filter and effaced by the IM relay. Consequently, the

IM relay on average expunges 0.8 % of instant messages with seven comprised character strings. The percentage rates of eradicable instant messages in Table 1 do not necessarily chime with those ones in real scenarios, but these percentages do not observably affect any clocked run durations and, due to this, lend themselves to a passable indication.

Table 1. Likelihood of instant messages with at least one obnoxious word

c (words per message)	1	2	3	4	5	6	7
P(c) (likelihood of obnoxiousness)	0.0011	0.0022	0.0034	0.0046	0.0057	0.0069	0.0080

5.2 Experimental Result

This subsection shall not merely reflect the measured data of the whitelisting-experiment but also compare them with the (querying) performance of Secure-String 3.0 [13], a safe sieve based on blacklisting. Because the experiment with SecureString 3.0 educed metered values as a function of the plaintext message length $|v|$, but the whitelisting-experiment extracted data as a function of the word count c, Table 2 lists the mean length of plaintext messages $|v|$ as a function of their word count c to enable a contrasting juxtaposition of both techniques.

Table 2. Average plaintext instant message lengths

c (words per message)	1	2	3	4	5	6	7
$|v|$ (mean characters per message)	8	16	25	34	43	51	60

Figure 6 visualizes for five methods the mean running time in nanoseconds of an instant message (inclusive its acknowledgment) between Alice and Bob as a function of its contained words c.

The red-colored data series emanates from whitelisting of plaintext instant messages. The renunciation of any encryption features the best performance, but it is strongly inadvisable due to too many security risks.

The obtainment of computational security by applying the safe whitelist screener based on ECC-192 [28] (light green-colored data series) slows the mean lead time down by factors between 22.46 ($c = 1$) and 10.80 ($c = 7$) in comparison to plaintext whitelisting.

If ECC-224 (green-colored data series) becomes exerted in place of plaintext whitelisting, the average processing time rises by factors between 24.79 ($c = 1$) and 11.95 ($c = 7$).

The highest deceleration with factors between 30.05 ($c = 1$) and 14.41 ($c = 7$) takes place if ECC-256 (dark green-colored data series) gets practiced instead of plaintext whitelisting.

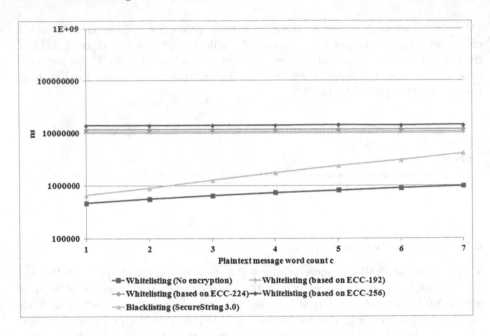

Fig. 6. Comparative experimental results (Color figure online)

The mean runtime of an instant message between Alice and Bob during the blacklisting-experiment with SecureString 3.0 (orange-colored data series) ranges between the red-colored values of plaintext whitelisting and the light green-colored ones of ciphertext whitelisting based on ECC-192. For an instant message with just one word ($c = 1$), the average transportation time of SecureString 3.0 (0.66 ms) resides nearby that one of plaintext whitelisting (0.47 ms), but it grows much faster with a rising word count than all other data series. With 4.09 ms for word count $c = 7$, SecureString 3.0 already expends approximately half as much time as the whitelisting based on ECC-192 with 10.61 ms. For still higher word counts, extrapolation of the orange-colored data series might show worse performance for SecureString 3.0 than for whitelisting based on ECC.

Avowedly, SecureString 3.0 can effectuate superior performance than safe whitelisting in case of short instant messages because of much fewer executed asymmetric cryptographic operations. Anyhow, the nearly impossible bypassing of whitelisting compared to the loopholes of blacklisting justifies all longer durations of safe whitelisting. Already Ferguson and Schneier taught that secure design philosophy forbids to cut a security corner in the name of efficiency, because there exist too many fast, insecure systems [15].

6   Conclusion

This disquisition walks the line to reconcile divergent expectations of IM lobbies. While the research community presses for enhancement of minors' media literacy and risk-awareness, parents often appreciate some censorship because of their too modest knowledge of online media and lack of time to trace their kids' activities. What is more, an IM architecture must be acceptably safe.

For that purpose, this disquisition manifests a whitelist-based IM screener for optional employment that particularly forfends IM consignees from being taken in by other IM lobbyists. Predominantly, this means the elimination of instant messages with inappropriate content before they can come through to their IM addressees. The contemplated recommendation rests upon the cooperation of state-of-the-art symmetrical and asymmetrical cryptosystems to evenhandedly accomplish the security objectives authenticity, integrity, privacy, and resilience. A decent security analysis corroborates the attainment of each security objective. In face of the developmental stage of the IM checker, an experiment with a condign prototype harvested valuable clues to its practicability.

Meaningful prospective continuations might refine the offered RMI-based IM sifter with a private ciphertext whitelist for each user as addition to the elaborated global one, because various users could interpret the same term either derogatorily or neutrally. Accessorily, future publications could take over or adapt the sifter utility at hand for manifold applications that need to screen ciphertext.

Acknowledgments. Many thanks to Bettina Baumgartner from the University of Vienna for proofreading this paper!

References

1. van den Berg, B.: Colouring inside the lines: using technology to regulate childrens behaviour online. In: van der Hof, S., van den Berg, B., Schermer, B. (eds.) Minding Minors Wandering the Web: Regulating Online Child Safety, Information Technology and Law Series, vol. 24, pp. 67–85. T.M.C. Asser Press, March 2014. doi:10.1007/978-94-6265-005-3_4

2. van den Berg, B.: Regulating online child safety: introduction. In: van der Hof, S., van den Berg, B., Schermer, B. (eds.) Minding Minors Wandering the Web: Regulating Online Child Safety, Information Technology and Law Series, vol. 24, pp. 1–16. T.M.C. Asser Press, March 2014. doi:10.1007/978-94-6265-005-3_1

3. Boss, G., McConnell, K.C.: System and method for filtering instant messages by context, US Patent App. 10/356,100. https://www.google.com/patents/US20040154022

4. Chai, Q., Gong, G.: Verifiable symmetric searchable encryption for semi-honest-but-curious cloud servers. In: 2012 IEEE International Conference on Communications (ICC), pp. 917–922, June 2012. doi:10.1109/ICC.2012.6364125

5. Ciavarella, R., de Terwangne, C.: Online social networks and young peoples privacy protection: the role of the right to be forgotten. In: van der Hof, S., van den Berg, B., Schermer, B. (eds.) Minding Minors Wandering the Web: Regulating Online Child Safety, Information Technology and Law Series, vol. 24, pp. 157–171. T.M.C. Asser Press, March 2014. doi:10.1007/978-94-6265-005-3_9

6. Daemen, J., Rijmen, V.: The Design of Rijndael: AES-the Advanced Encryption Standard. Springer, New York (2001)

7. Fahrnberger, G.: Computing on encrypted character strings in clouds. In: Hota, C., Srimani, P.K. (eds.) Distributed Computing and Internet Technology. LNCS, vol. 7753, pp. 244–254. Springer, Heidelberg (2013). doi:10.1007/978-3-642-36071-8_19

8. Fahrnberger, G.: Securestring 2.0 - a cryptosystem for computing on encrypted character strings in clouds. In: Eichler, G., Gumzej, R. (eds.) Networked Information Systems. Fortschritt-Berichte Reihe 10, vol. 826, pp. 226–240. VDI Düsseldorf, June 2013. doi:10.13140/RG.2.1.4846.7521/3

9. Fahrnberger, G.: A second view on securestring 2.0. In: Natarajan, R. (ed.) ICDCIT 2014. LNCS, vol. 8337, pp. 239–250. Springer, Heidelberg (2014). doi:10.1007/978-3-319-04483-5_25

10. Fahrnberger, G.: SIMS: a comprehensive approach for a secure instant messaging sifter. In: 2014 IEEE 13th International Conference on Trust, Security and Privacy in Computing and Communications (TrustCom), pp. 164–173, September 2014. doi:10.1109/TrustCom.2014.25

11. Fahrnberger, G.: Repetition pattern attack on multi-word-containing securestring 2.0 objects. In: Natarajan, R., Barua, G., Patra, M.R. (eds.) ICDCIT 2015. LNCS, vol. 8956, pp. 265–277. Springer, Heidelberg (2015). doi:10.1007/978-3-319-14977-6_26

12. Fahrnberger, G.: A detailed view on securestring 3.0. In: Chakrabarti, A., Sharma, N., Balas, V.E. (eds.) Advances in Computing Applications. Springer, Singapore, November 2016. doi:10.1007/978-981-10-2630-0_7

13. Fahrnberger, G., Heneis, K.: Securestring 3.0 - a cryptosystem for blind computing on encrypted character strings. In: Natarajan, R., Barua, G., Patra, M.R. (eds.) ICDCIT 2015. LNCS, vol. 8956, pp. 331–334. Springer, Heidelberg (2015). doi:10.1007/978-3-319-14977-6_33

14. Fahrnberger, G., Nayak, D., Martha, V.S., Ramaswamy, S.: Safechat: a tool to shield children's communication from explicit messages. In: 2014 14th International Conference on Innovations for Community Services (I4CS), pp. 80–86, June 2014. doi:10.1109/I4CS.2014.6860557

15. Ferguson, N., Schneier, B.: Practical Cryptography. Wiley, New York (2003)

16. Ganz, H., Borst, K.J.: Multiple-layer chat filter system and method, US Patent 8,316,097. https://www.google.com/patents/US8316097

17. Genner, S.: Violent video games and cyberbullying: why education is better than regulation. In: van der Hof, S., van den Berg, B., Schermer, B. (eds.) Minding Minors Wandering the Web: Regulating Online Child Safety, Information Technology and Law Series, vol. 24, pp. 229–243. T.M.C. Asser Press, March 2014. doi:10.1007/978-94-6265-005-3_13

18. van der Hof, S.: No childs play: online data protection for children. In: van der Hof, S., van den Berg, B., Schermer, B. (eds.) Minding Minors Wandering the Web: Regulating Online Child Safety, Information Technology and Law Series, vol. 24, pp. 127–141. T.M.C. Asser Press, March 2014. doi:10.1007/978-94-6265-005-3_7

19. ISO, IEC 10116:2006: Information technology - security techniques - modes of operation for an n-bit block cipher. International Organization for Standardization, February 2006. https://www.iso.org/iso/catalogue_detail?csnumber=38761

20. van der Knaap, L.M., Cuijpers, C.: Regulating online sexual solicitation: towards evidence-based policy and regulation. In: van der Hof, S., van den Berg, B., Schermer, B. (eds.) Minding Minors Wandering the Web: Regulating Online Child Safety, Information Technology and Law Series, vol. 24, pp. 265–281. T.M.C. Asser Press, March 2014. doi:10.1007/978-94-6265-005-3_15

21. Landsman, R.: Filter for instant messaging, US Patent Ap. 11/252,664. https://www.google.com/patents/US20070088793

22. Lievens, E.: Children and peer-to-peer risks in social networks: regulating, empowering or a little bit of both? In: van der Hof, S., van den Berg, B., Schermer, B. (eds.) Minding Minors Wandering the Web: Regulating Online Child Safety, Information Technology and Law Series, vol. 24, pp. 191–209. T.M.C. Asser Press, March 2014. doi:10.1007/978-94-6265-005-3_11
23. Liu, Z., Lin, W., Li, N., Lee, D.: Detecting and filtering instant messaging spam - a global and personalized approach. In: 1st IEEE ICNP Workshop on Secure Network Protocols, (NPSec), pp. 19–24, November 2005. doi:10.1109/NPSEC.2005.1532048
24. Miller, G.A.: Wordnet: a lexical database for english. Commun. ACM **38**(11), 39–41 (1995). doi:10.1145/219717.219748
25. Notten, N.: Taking risks on the world wide web: the impact of families and societies on adolescents risky online behavior. In: van der Hof, S., van den Berg, B., Schermer, B. (eds.) Minding Minors Wandering the Web: Regulating Online Child Safety, Information Technology and Law Series, vol. 24, pp. 105–123. T.M.C. Asser Press, March 2014. doi:10.1007/978-94-6265-005-3_6
26. Örencik, C., Savas, E.: An efficient privacy-preserving multi-keyword search over encrypted cloud data with ranking. Distrib. Parallel Databases **32**(1), 119–160 (2014). doi:10.1007/s10619-013-7123-9
27. Sonck, N., de Haan, J.: Safety by literacy? rethinking the role of digital skills in improving online safety. In: van der Hof, S., van den Berg, B., Schermer, B. (eds.) Minding Minors Wandering the Web: Regulating Online Child Safety, Information Technology and Law Series, vol. 24, pp. 89–104. T.M.C. Asser Press, March 2014. doi:10.1007/978-94-6265-005-3_5
28. Stebila, D., Green, J.: Elliptic curve algorithm integration in the secure shell transport layer. RFC 5656 (Proposed Standard), December 2009. https://www.ietf.org/rfc/rfc5656.txt
29. Thierer, A.: A framework for responding to online safety risks In: van der Hof, S., van den Berg, B., Schermer, B. (eds.) Minding Minors Wandering the Web: Regulating Online Child Safety, Information Technology and Law Series, vol. 24, pp. 39–66. T.M.C. Asser Press, March 2014. doi:10.1007/978-94-6265-005-3_3
30. Vandebosch, H.: Addressing cyberbullying using a multi-stakeholder approach: the flemish case. In: van der Hof, S., van den Berg, B., Schermer, B. (eds.) Minding Minors Wandering the Web: Regulating Online Child Safety, Information Technology and Law Series, vol. 24, pp. 245–262. T.M.C. Asser Press, March 2014. doi:10.1007/978-94-6265-005-3_14
31. Waksman, A., Sethumadhavan, S.: Tamper evident microprocessors. In: 2010 IEEE Symposium on Security and Privacy (SP), pp. 173–188, May 2010. doi:10.1109/SP.2010.19
32. van der Zwaan, J.M., Dignum, V., Jonker, C.M., van der Hof, S.: On technology against cyberbullying. In: van der Hof, S., van den Berg, B., Schermer, B. (eds.) Minding Minors Wandering the Web: Regulating Online Child Safety, Information Technology and Law Series, vol. 24, pp. 211–228. T.M.C. Asser Press, March 2014. doi:10.1007/978-94-6265-005-3_12

Collaboration and Workflow

Adaptive Workflow System Concept for Scientific Project Collaboration

Vasilii Ganishev[✉], Olga Fengler, and Wolfgang Fengler

Computer Science and Embedded Systems Group,
Ilmenau University of Technology, 98693 Ilmenau, Germany
{vasily.ganishev,olga.fengler,wolfgang.fengler}@tu-ilmenau.de

Abstract. Scientific projects are dynamic processes, and possibilities to determine their structure beforehand is matter of great practical importance. Some communities that work on the same project can join scientists from various countries and research centers. The approaches related to BPM (Business Process Management) are rigid and are "process–centric", therefore they provide limited flexibility for some of process handling requirements. ACM (Advanced Case Management), in contrary, provides the necessary degree of flexibility. Moreover its synthesis with the BPM suggested attractive advantages for process configuration and control.

In this paper we propose a concept of adaptive workflow system for scientific project collaboration and introduce the description of each part in detail. A user describes a project in terms of cases. Then system suggests suitable possible process models, that the user is free to follow or not. At further steps the system specifies its suggestions. This approach combines the flexibility of ACM with clearness of BPM.

Keywords: Workflow management systems · Collaboration systems · Scientific project collaboration · Adaptive case management

1 Introduction

Nowadays the internet and the development of other communication facilities have led the rise of great scientific collaborations. Unlike "Big Science" these collaborations have much more flexible structure and are being characterized by possible geographical diversity of collaborators [3]. Some communities that work on the same project can join scientists from various countries and research centres. In this case there is a need in tools to lighten a team-work and to make collaboration more clear. However, it should not be understood as another scientific workflow system such as Kepler, Taverna or FireWorks which help to control research project from the task concurrency point of view. This system should configure, support and optimize collaborative work of a group of individual scientists. Therefore, a high degree of adaptivity is required for handling and structuring the processes beforehand. One of the practical necessities for developing such systems is their use in settling the workflow of the scientific projects which are known to be highly dynamic and long–term processes.

G. Fahrnberger et al. (Eds.): I4CS 2016, CCIS 648, pp. 115–128, 2016.
DOI: 10.1007/978-3-319-49466-1_8

For a long time there is a discussion about lack of proper degree of adaptivity in workflow systems. Rigid Business Process Management (rigid BPM) does not provide necessary process evolution and/or exception handling mechanisms. If one takes into consideration all possible changes in the process, it makes process presentation unreadable and sophisticated which eliminates almost all advantages of such an approach [2].

Some researchers propose to use high-level change operations to provide structural adaptivity such as "Insert Process Fragment" [17] or with change primitives higher degree of granularity like "Add Node" and "Add Arc" in recovery nets [9]. These approaches require the use of change correctness and change control systems (e.g. Opera [8]). Despite several advantages, these concepts require complex process support.

In our opinion the problem lies in the basic concept of BPM, which is too restrictive and has problems dealing with change [2]. Complex projects, not only limited to scientific projects, require collaboration of knowledge workers, whose work may not fit predefined process model due to changing circumstances (new information, feedback from already done steps, exceptions, etc.). The BPM, more suitable for short routine operations, does not provide possibilities to handle this, because it is too much "process–centric".

Advanced (or adaptive) case management (ACM) is relatively new concept to configure knowledge work processes. It is more "goal–centric" and defines the approach as "what could be done", rather than "how should it be done" [16]. This concept provides more flexible process description, however it has almost lost the connection with BPM and acts completely separate. Therefore, we propose that the ACM together with the BMP could make a great contribution to adaptive workflow systems. In this paper we will mainly address to this problem, filling the gap between the flexibility of the ACM and the accuracy of the BPM. The article is organized as follows: in the second section the original concept of ACM is described. The third section proposes an idea about the concept of adaptive workflow system for scientific project collaboration, its subsections describe each part of the system in detail. In fourth section we will show the operation of the proposed system on the example project. In last section we will make conclusions and present further directions of the research.

2   Advanced Case Management

Case management was used to support decision-making processes such as patient predictors of improvement [13] in 1990s. First significant attempts to leave "process–centric" ideology in favour of "case–centric" were made in the middle 2000s [2,7]. In 2010 the term Advanced Case Management was introduced in the field of process management by Workflow Management Coalition. However, for a long time there was no accurate notation for ACM till Object Management Group (OMG) proposed Case Management Model and Notation v1.0 in 2014 [6].

ACM considers processes from the point of view of cases. A case describes a goal of processing and is a set of information (as documents, accepted production practices and decisions) that leads to this goal. Each case can be resolved in ad-hoc manner completely. For some similar cases it is possible to figure out a set of common practices following to optimal solution. ACM helps knowledge workers to undertake these practices for tasks, that lack high degree of adaptivity. Unlike rigid BPM the ACM allows case workers to plan and change a set of tasks for a case, to change task order and to collaborate with other knowledge workers during task execution.

Case management presumes two phases of the process modeling: design-time and run-time. In the design-time phase the designer (usually with expertise in business analytics) predetermines a set of tasks that is necessary to achieve the case goal and possibly some non-mandatory (in the CMMN standard it is called discretionary [6]) tasks that are available to a knowledge worker. During the run-time phase case workers execute tasks as planned and possibly add some new discretionary tasks that were not supposed initially.

Case itself consists of a Plan Model and a set of Case Roles that is needed for executor assignment. The Plan Model describes case behaviour during its execution and is depicted as a folder shape. The Plan Model is used as a container for Plan Items and Case Files that are representing useful information for Case resolution. CMMN predetermines four types of Plan Items: Task, Stage and Milestone. Description of these Plan Items is as follows:

2.1 Task

Task represents a unit of work. Tasks are depicted by rectangle shapes with rounded corners. There are four types of Tasks:

Blocking Human Task is a type of task, that is performed by a knowledge worker and waits until a piece of work associated with it completes (e.g. Decision). This type is represented with human icon in the upper left corner of a task shape.

Non-Blocking Human Task is a type of task, that is performed by a knowledge worker and is considered as completed immediately when it is opened (e.g. Review). This type is represented with hand icon in the upper left corner of a task shape.

Process Task is a type of task, that is performed by a script or business process. It performs without human interaction. This type is represented with chevron icon in the upper left corner of a task shape.

Case Task is a type of task, that is linked with another Case. It helps to organize Case hierarchy. This type is represented with folder icon in the upper left corner of a task shape.

Tasks could be both mandatory and discretionary. Discretionary tasks are represented by rectangle shapes with dashed lines and rounded corners.

2.2 Stage

A Stage is a container for Tasks that constitute a separate episode of the case. Stages can be regarded as sub-cases in analogy with sub-processes in BPM, similar to sub-processes they make models clearer by hiding complex behaviour behind abstract steps. Stage is depicted as a rectangle shape with angled corners and a marker '+' or '-' at bottom centre, which shows its expended or collapsed state. In expended state all tasks in stage are shown, while in collapse state all are hidden. Stages also could be both mandatory and discretionary. When discretionary stages are represented by rectangle shapes with dashed lines and angled corners.

2.3 Milestone

A Milestone is a sub-target of the Case, defined to enable evolution of Case resolution progress. Milestone is associated with no specific piece of work, but it is considered as achieved when associated tasks are completed or a Case File becomes available. Milestone is represented by a rectangle shape with half-rounded ends.

There are two more essentials that help to organize semantic dependencies between Plan Items: Entry and Exit Criteria (in CMMN they are called Sentries [6]) and Connector that helps to link Plan Items together.

2.4 Sentries and Connectors

Sentries define criteria by which Plan Items become enabled (entry criterion) or terminate (exit criterion). Sentry with entry criterion is represented as an empty diamond, and Sentry with exit criterion – as a solid diamond.

A criterion in a Sentry is a combination of an event (trigger, On-Part) and/or a condition (if-expression). If each existing criterion part evaluates to "true", then Sentry is satisfied. Satisfaction of a Sentry with entry criterion will follow to either enabling of a Task or a Stage or to an achievement of a Milestone, depending on what it was associated with. By triggering a Sentry with exit criterion associated Task or Stage will be terminated.

Connectors are used to visualize dependencies among Plan Items, e.g. On-Part of a Sentry. Connectors are graphically represented in notation as dotted lines.

3 Concept of System

In this section we propose the concept of adaptive workflow system for scientific project collaboration. Since, scientific projects are knowledge works, so it is hard to determine exact process flow and even harder to follow all its constraints. During execution the model of process will define itself more accurately and can be extended by exception handling, if there occurs any exception. The architecture of proposed system looks as follows (Fig. 1), at first we describe the concept from the point of view of different user roles and system parts.

Manager (Process Owner). A Manager is the main actor in the system. This actor is responsible for determination of scientific project primary goal, setting

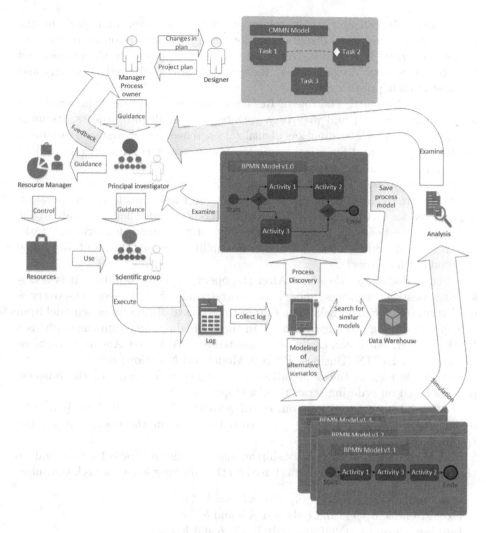

Fig. 1. Architecture of adaptive workflow system

the initial project schedule and budget. Manager also approves important decisions as the transition between design-time and run-time stages or primary goal changes.

Designer. As the name implies, a Designer plays a crucial role at the design-time phase. The main task of the designer is to determine a set of Plan Items and dependencies among them required to achieve the project primary goal. The project's plan in the form of CMMN diagram should be approved by a Manager and then present itself as instructions for project execution. When major changes in the project accrue, the designer could be called to review the project plan.

Principal Investigator. This role represents "mid–level manager" of the project. Its mission is to distribute tasks among scientific groups, to monitor compliance with the schedule, to manage provided resources in the project and to take operational decisions (e.g. inclusion of new additional sub-goals and defining of their priorities).

Resource Manager. The role of Resource Manager is optional (in sense that it can be automated) and may be associated with multiple projects. Its main responsibilities are: coordination of utilization schedules of material resources (such as experimental facilities) and human resources (scientific groups and/or single scientists) among several projects.

Scientific Group. This role is a main working unit of the project. Scientific groups report to the system on the work done, so these data form an operation log required for further analysis. Users in this group decide themselves about the order in which tasks are solved, the necessity of predefined discretionary tasks execution and adding new tasks (this possibility can be ruled out by policy adopted on the project).

Process discovery subsystem. After the operation log is obtained it is necessary to construct a process model that corresponds to it. Process Discovery is used very often for such tasks. The simplest method to obtain process model from the operational log is α–algorithm [1]. In the original α–algorithm the model is a WorkFlow-net, but it can easily be translated to YAWL (Yet Another Workflow Language) or BPMN (Business Process Model and Notation) nets.

The basic idea of the α–algorithm is to define a footprint of the business process based on ordering strict relationship:

Let a and b be activities from set of activities A. $a \prec b$ if there is a trace $\sigma = t_1, t_2, \cdots, t_n$ and $i \in [1, \cdots, n]$ such that σ is in the operation log and $t_i = a, t_{i+1} = b$.

Based on this ordering relationship among activities it is possible to introduce the concept of the footprint that is based on the following log–order relationships:

- Strict order. $a \rightarrow b$ if and only if $a \prec b$ and $b \nprec a$
- Exclusiveness. $a \# b$ if and only if $a \nprec b$ and $b \nprec a$
- Interleaving order. $a \| b$ if and only if $a \prec b$ and $b \prec a$

But the α–algorithm is suitable for repeating processes otherwise it is impossible to determine parallel activities. In the case of scientific project the process executes only once.

In order to bypass this restriction it is necessary to report about one activity twice: at the beginning of execution (it could be automated, e.g. assigned and active task means the beginning of its execution) and at the end. After that the log determines extended set of activities $A^* : (a'_1, \cdots, a'_n) \cup (a^*_1, \cdots, a^*_n)$, where a'_1 is the beginning of the activity a_1 and a^*_1 — its end. So order log–order relations can be overwritten:

- Strict order. $a \rightarrow b$ if and only if $a^* \prec b'$
- Interleaving order. $a \| b$ if and only if $a' \prec b'$ and $b' \prec a^*$

The simple α-algorithm is defined as follows (instead of WorkFlow net semantics we use YAWL semantics) [1]:

Let L be an event log over some subset T of a set of activities A. Activities from T define the activities in a generated YAWL net. T_I is the start activity, that appears first in the log L. T_O is the end activity, that appears last in the log L. Next two steps are required to discover a control flow F that defines an order of activities:

$$X = \{(I,O) \mid I \subseteq T \cap I \neq \emptyset \cap O \subseteq T \cap O \neq \emptyset \cap \forall_{i \in I} \forall_{o \in O} i \rightarrow o \cap \forall_{i_1,i_2 \in I} i_1 \parallel i_2 \cap \forall_{o_1,o_2 \in O} o_1 \parallel o_2\}$$
$$F = \{(i,o) \mid i \in I \cap o \in O\}$$

The resulting YAWL net is a tuple $\alpha(L) = (T,F)$. It is worth noting that the resulting net has a representational bias it supports cycles only in their unrolled form.

But until the research project is ended this model is not sound, it defines only the part of the process. The control flow for future activities cannot be exactly predicted, moreover future activities themselves may be unknown. But the system can assume the rest of the process, this is the function of the next subsystem.

Simulation and analytics subsystem. CMMN model predetermine not a single process, but a set of business processes that do not break semantic dependencies of the plan. In other words the CMMN model defines a partially ordered set T of tasks. It is also worth noting that a task in CMMN model does not correspond to a single action in BPM notations such as BPMN or YAWL, but it can be also whole process fragments. Nevertheless such information may be useful if not for current project, then in the next or in analysing the difficulties during the project.

By opening the notion of each task, one can obtain a partially ordered set of activities A. It is possible, when some activities to solve it are done or business process for it is defined. Tasks those do not have yet corresponding activities may be considered as subprocesses. This is consistent with the statement that "...the control structure a business process can be seen as a partially ordered set of activities..." [12]. So a set of all possible business processes B obtained from this model is a set of *linear extensions* Ω of partially ordered set A.

A binary relation φ defined on set M by certain set R_φ is called partial order, if it possesses following properties [15]:

1. Reflexivity: $\forall a(a \varphi a)$
2. Transitivity: $\forall a, b, c(a \varphi b) \wedge (b \varphi c) \Rightarrow (a \varphi c)$
3. Antisymmetry: $\forall a, b(a \varphi b) \wedge (b \varphi a) \Rightarrow a = c$

And the set M (or rather the pair (M, φ)) is called the partially ordered set or poset. Usually the partial order relation is denoted by $\preceq$ symbol.

It is worth noting that activities in business process do not always satisfy the partial order: activity b follows activity a and activity a follows activity b do not mean that activities a and b are equal, but they are parallel, it is so called interleaving order relation [11]. In this case, for simplicity, one can consider during generation that between these activities there is no order relation at all.

Under the permutation we mean the act of arranging the set A. Thus a linear extension δ of the poset A is a permutation of the elements where $\delta(i) \preceq \delta(j)$ implies $i \preceq j$ in the original set A [10].

For the solution of this problem one can find several methods in literature, e.g. [10,14].

In [14] authors generated all possible linear extensions of partially ordered set of positive numbers (but this approach works on any set with a given partial order) through family tree traversal. Family tree is a special type of tree, where node and leaves are permutation of a given partially ordered set, the connections represent parent-child relations between two permutations.

After obtaining a set of possible business processes, it is possible to count for each some useful parameters as consistency with the project schedule, costs, etc. and choose an optimal business process model at this step. It should be noted that till there corresponding activities for all tasks defined in CMMN model, business process model has an approximate character. Because of that the optimal business process model should be recalculated after every process step and in general results may vary. Also note that the activities among which the partial order is not defined can be executed in parallel.

Process Data Warehouse. Some processes or their parts can follow to the same results, create similar repeating pattern. Because of that it is worth to save successful processes, also the processes, which ended with fault. It is required for analysis as well as to avoid such situations in the future. The corresponding system module provides a report to principal investigator.

One must specify a measure of similarity between two business processes. Under the measure of similarity between two business processes we mean a real-valued function *dist* defined on the set of business processes $\mathbb{B}$ that quantifies the similarity between these processes. This function should meet the following requirements [4]:

1. Nonnegativity: $\forall B_0, B_1 \in \mathbb{B} \; dist(B_0, B_1) \geq 0$
2. Symmetry: $\forall B_0, B_1 \in \mathbb{B} \; dist(B_0, B_1) = dist(B_1, B_0)$
3. Equality: $\forall B_0, B_1 \in \mathbb{B} \; dist(B_0, B_1) = 0 \Leftrightarrow B_0 \equiv B_1$
4. Triangle inequality: $\forall B_0, B_1, B_2 \in \mathbb{B} \; dist(B_0, B_2) \leq dist(B_0, B_1) + dist(B_1, B_2)$

This topic has been well studied. In literature four specific classes of such measures are identified [4]:

1. Measures based on the correspondence between nodes and edges in the process model.
   These metrics are calculated using syntactic or/and semantic similarity among activities, labels, etc. in two models.
2. Measures based on the edit distance between graphs.
   These approaches determine the similarity of model structures.
3. Approaches based on the sets of traces.
   These metrics compare the behaviour of models experimentally by simulating its operation.

4. Measures based on causal dependencies between activities.
   These metrics compare the behaviour of models analytically.

The first class of measures helps to find related models with same components. The second metric type analyses model structure, while it is defined clearly only at the end of functioning. Approaches in the third class generate possible traces of the process model, which can cause rapid increase in the number of sequences that require analysis as well as potential occurrence of infinite sequences in the case of cycles. The fourth class of metrics is suitable for use in this case.

One measure from the fourth class is m^3 similarity measure [11] that uses behavioural profile of the process for analysis. The process discovery subsystem defines footprint that is similar to behavioural profile of the process during its execution (See p. 6), but the behavioural profile is based on the other concept of ordering relationship:

Let a and b be activities from set of activities A. $a \preceq b$ if there is a trace $delta = t_1, t_2, \cdots, t_n$ and $i \in [1, \cdots, n], j \in [1, \cdots, n], i < j$ such that $delta$ is in the operation log and $t_i = a, t_j = b$.

The log–order relationships are as follows:

- Strict order. $a \rightarrow b$ if and only if $a \preceq b$ and $b \npreceq a$
- Exclusiveness. $a\#b$ if and only if $a \npreceq b$ and $b \npreceq a$
- Interleaving order. $a\|b$ if and only if $a \preceq b$ and $b \preceq a$

To determine the degree of similarity it is necessary to calculate three similarity metrics [11]:

Exclusiveness Similarity.
$$s_+(B_0, B_1) = \frac{|+_{B_0} \cap +_{B_1}|}{|+_{B_0} \cup +_{B_1}|}$$
Strict Order Similarity.
$$s_\rightarrow(B_0, B_1) = \frac{|\rightarrow_{B_0} \cap \rightarrow_{B_1}|}{|\rightarrow_{B_0} \cup \rightarrow_{B_1}|}$$
Interleaving Order Similarity.
$$s_\|(B_0, B_1) = \frac{1}{2} \cdot \frac{|(\rightarrow_{B_0} \cup \|_{B_0}) \cap \rightarrow_{B_1}|}{|\rightarrow_{B_0} \cup \|_{B_0} \cup \rightarrow_{B_1}|} + \frac{|\rightarrow_{B_0} \cap (\rightarrow_{B_1} \cup \|_{B_1})|}{|\rightarrow_{B_0} \cup \rightarrow_{B_1} \cup \|_{B_1}|}$$

Then the m^3 similarity measure is defined as a remainder of subtracting the weighted sum of the previous three metrics from one [11]:
$$m^3(B_0, B_1) = 1 - \sum_i w_i \cdot s_i(B_0, B_1)$$

with $i \in \{+, \rightarrow, \|\}$ and weighting factors w_i such that $\sum_i w_i = 1$.

While the m^3 measure satisfies the property of triangle inequality, it is possible to bound the similarity of two models with already computed distances [11]:
$$dist(B_0, B_2) \leq |dist(B_0, B_1) + dist(B_1, B_2)|$$
$$dist(B_0, B_2) \geq |dist(B_0, B_1) - dist(B_1, B_2)|$$
This reduces the computation by rejecting in advance not prospective models.

4   Example

In this section the operation of the process discovery subsystem and the subsystem of simulation and analytics on the example of speaker clustering implementation project is discussed [5]. Speaker clustering is a task of division of voice recordings into classes so that in each class there were only records owned by one user. Each record contains the voice of a single user. Most often, the clustering process takes place without direct supervision, a priori knowledge of the number of classes or their structure. This process is often an integral part of speaker recognition or speech recognition tasks.

The CMMN model shows the plan of the project that was approved (Fig. 2).

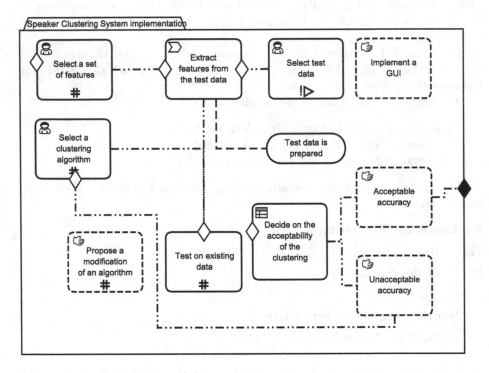

Fig. 2. CMMN model of speaker clustering implementation project

The scientific group should select a clustering algorithm ("Select a clustering algorithm" task). The pure voice signal is difficult to analyse, because it contains a large amount of information noise. But it is possible to extract some specific features from it and use them as input vector for clustering ("Select a set of features" task). Also the test data should be selected, it should be marked to simplify the algorithm testing ("Select test data" task). All these three tasks can be executed in parallel, because there is no data dependence among them. Once "Select a set of features" and "Select test data" task are accomplished a set of features can be extracted from the test data ("Extract features from the test data" task) it immediately follows to the achievement of "Test data is prepared" milestone that describes the sub-goal of the project. After this the task "Test on existing data" is enabled. After its executing the decision should be taken: is the resulting accuracy acceptable or not. If it is acceptable the case closes, else some tasks or all of them can be repeated. The primary plan also contains two optional discretionary tasks: "Implement a GUI" and "Propose a modification of an algorithm".

For example, at some project phase the log contains following recordings (Table 1). The column "Activity" shows the name of activity, an apostrophe after the activity name means the beginning of its execution, two apostrophes mean the end. The column "Case task" present the name of the corresponding task in CMMN model (Fig. 2), "None" means that this task was not planned beforehand. The column "Timestamp" shows the time, when the event connected with activity happened in format "yyyy-mm-ddThh:mm".

Table 1. Log of speaker clustering project

Activity	Case task	Timestamp
AHC study'	Select a clustering algorithm	2014-10-27T09:32+03:00
Lyon model study'	Select a set of features	2014-10-28T10:07+03:00
Search for test data'	Select test data	2014-10-28T12:14+03:00
Lyon model study"	Select a set of features	2014-10-30T15:24+03:00
Search for test data"	Select test data	2014-11-03T16:21+03:00
PLPC study'	Select a set of features	2014-11-04T08:19+03:00
PLPC study"	Select a set of features	2014-11-07T11:47+03:00
LFCC study'	Select a set of features	2014-11-07T13:55+03:00
LFCC study"	Select a set of features	2014-11-11T12:03+03:00
MFCC study'	Select a set of features	2014-11-17T08:44+03:00
MFCC study"	Select a set of features	2014-11-21T19:42+03:00
MFCC extraction'	Extract features from the test data	2014-12-03T15:14+03:00
AHC study"	Select a clustering algorithm	2014-12-05T10:15+03:00
AHC implementation'	Select a clustering algorithm	2014-12-12T12:35+03:00
MFCC extraction"	Extract features from the test data	2014-12-22T17:03+03:00
AHC implementation"	Select a clustering algorithm	2015-01-07T12:27+03:00
Test1'	Test on existing data	2015-01-07T12:32+03:00
Test1"	Test on existing data	2015-01-07T16:20+03:00
Calculate accuracy'	Decide on the acceptability of the clustering	2015-01-07T16:31+03:00
Calculate accuracy"	Decide on the acceptability of the clustering	2015-01-07T16:35+03:00
Unacceptable accuracy"	Unacceptable accuracy	2015-01-07T16:37+03:00
ALIZE study'	None	2015-01-14T12:07+03:00
ALIZE study"	None	2015-01-20T11:27+03:00
SOINN study'	Select a clustering algorithm	2015-01-20T13:03+03:00

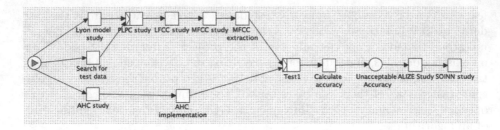

Fig. 3. YAWL net built on the log

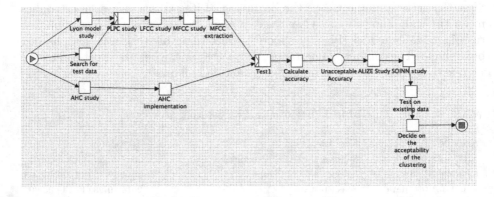

Fig. 4. One of generated YAWL nets

The process discovery subsystem translates the log into the table presented in Table 2, describing partial order among activities. According to the results given in the table the resulting YAWL net can be represented in Fig. 3.

Table 2. Partial order among activities

Name Name	AHC study	Lyon model study	Search for test data	PLPC study	LFCC study	MFCC study	MFCC extraction	AHC implementation	Test1	Calculate accuracy	Unacceptable accuracy	ALIZE study	SOINN study
AHC study		‖	‖	‖	‖	‖	‖	→					
Lyon model study	‖		‖	→									
Search for test data	‖	‖		→									
PLPC study	‖	←	←			→							
LFCC study	‖			←		→							
MFCC study	‖				←		→						
MFCC extraction	‖					←		‖	→				
AHC implementation	←						‖		→				
Test1						←		←		→			
Calculate accuracy									←		→		
Unacceptable accuracy										←		→	
ALIZE study											←		→
SOINN study												←	

The resulting net is not sound, because there is no output condition, which means that the project is still in progress.

The subsystem of simulation and analytics will extend the set of activities from log with active tasks from CMMN model. All tasks are now active except "Select a clustering algorithm", because it is executed second time and its condition of repetition is not met. After that all possible linear extensions will be generated, such as given in Fig. 4.

After the project ends its YAWL model is saved in process data warehouse and is used in the next project, if the same patterns in the whole process or its parts are discovered.

5 Summary and Conclusion

Complex long–term scientific projects require more flexibility than it is offered by Business Process Management tools. The concept of Advanced Case Management make it possible to introduce the project in the form of an imprecise plan. However, without introduction of clarity at every stage, it is impractical to manage or improve the executable process. That is why ACM together with BPM can give more advantages to process configuration and control.

In this paper the system concept for scientific project collaboration is proposed. It supports the scientific project execution, helps to identify the best practices or avoid faults that occurred in the past. This involves the separation of actors or participants by roles. By proposed separation some roles can be combined in smaller scientific groups.

The operation of the two subsystems was shown on the example of speaker clustering project.

The analytical part of the system consists of three modules:

1. Process discovery subsystem allows to convert the log of activities to the business process model, which is required for further analysis.
2. Subsystem of simulation and analytics generates possible execution scenarios and analyses them according to certain criteria.
3. Process data warehouse is necessary to keep the information about previous projects and analyse similarity of current project to them. This can be useful to avoid previous faults as well as adhere to established practices.

In future work we will address the mechanism of obtaining a log of work as well as criteria for selecting the most appropriate process of the generated ones. Also further investigations in the consideration of human resources performance with fuzzy logic or the use of defeasible logic in describing the semantic relationships among tasks in the CMMN model are planned.

References

1. van der Aalst, W.M.P.: Process Discovery: An Introduction. Process Mining: Discovery, Conformance and Enhancement of Business, pp. 125–156. Springer, Heidelberg (2011)

2. van der Aalst, W.M., Weske, M., Gruenbauer, D.: Case handling: a new paradigm for business process support. Data Knowl. Eng. **53**(2), 129–162 (2005)

3. Beaver, D.D.: Reflections on scientific collaboration (and its study): past, present, and future. Scientometrics **52**(3), 365–377 (2005). http://dx.doi.org/10.1023/A:1014254214337

4. Becker, M., Laue, R.: A comparative survey of business process similarity measures. Comput. Ind. **63**(2), 148–167 (2012)

5. Ganishev, V., Vagin, V.: Speaker clustering using enhanced self-organizing incremental neural networks. Programmnye produkty i sistemy (Softw. Syst.) **3**, 136–142 (2015)

6. Group, O.M: Case Management Model and Notation (2014). http://www.omg.org/spec/CMMN/1.0/

7. Guenther, C., Reichert, M., van der Aalst, W.: Supporting flexible processes with adaptive workflow and case handling. In: 2008 IEEE 17th Workshop on Enabling Technologies: Infrastructure for Collaborative Enterprises, WETICE 2008, pp. 229–234, June 2008

8. Hagen, C., Alonso, G.: Exception handling in workflow management systems. IEEE Trans. Softw. Eng. **26**(10), 943–958 (2000). http://dx.doi.org/10.1109/32.879818

9. Hamadi, R., Benatallah, B.: Recovery nets: towards self-adaptive workflow systems. In: Zhou, X., Su, S., Papazoglou, M.P., Orlowska, M.E., Jeffery, K. (eds.) WISE 2004. LNCS, vol. 3306, pp. 439–453. Springer, Heidelberg (2004). doi:10.1007/978-3-540-30480-7_46

10. Huber, M.: Near-linear time simulation of linear extensions of a height-2 poset with bounded interaction. Chicago. J. Theor. Comput. Sci. **3**, 1–16 (2014)

11. Kunze, M., Weidlich, M., Weske, M.: m3-a behavioral similarity metric for business processes. Ser. Komposition (2011)

12. Moldt, D., Valk, R.: Object oriented petri nets in business process modeling. In: Aalst, W., Desel, J., Oberweis, A. (eds.) Business Process Management: Models, Techniques, and Empirical Studies, vol. 1806, pp. 254–273. Springer, Heidelberg (2000)

13. Mueser, K.T., Bond, G.R., Drake, R.E., Resnick, S.G.: Models of community care for severe mental illness: a review of research on case management. Schizophr. Bull. **24**(1), 37–74 (1998). http://schizophreniabulletin.oxfordjournals.org/content/24/1/37.abstract

14. Ono, A., Nakano, S.: Constant time generation of linear extensions. In: Kosowski, A., Walukiewicz, I. (eds.) FCT 2015. LNCS, vol. 9210, pp. 445–453. Springer, Heidelberg (2005). doi:10.1007/11537311_39

15. Simovici, D., Djeraba, C.: Mathematical Tools for Data Mining: Set Theory, Partial Orders, Combinatorics, chap. Springer, London (2008)

16. Swenson, K.D., Palmer, N., Silver, B.: Taming the Unpredictable Real World Adaptive Case Management: Case Studies and Practical Guidance. Future Strategies, October 2011

17. Weber, B., Reichert, M., Rinderle-Ma, S.: Change patterns and change support features enhancing flexibility in process-aware information systems. Data Knowl. Eng. **66**(3), 438–466 (2008). http://www.sciencedirect.com/science/article/pii/S0169023X0800058X

Zebras and Lions: Better Incident Handling Through Improved Cooperation

Martin Gilje Jaatun[1(✉)], Maria Bartnes[2], and Inger Anne Tøndel[1]

[1] Department of Software Engineering, Safety and Security, SINTEF ICT, Trondheim, Norway
{Martin.G.Jaatun,Inger.A.Tondel}@sintef.no
[2] SINTEF Group Head Office, Trondheim, Norway
Maria.Bartnes@sintef.no

Abstract. The ability to appropriately prepare for, and respond to, information security incidents, is of paramount importance, as it is impossible to prevent all possible incidents from occurring. Current trends show that the power and automation industry is an attractive target for hackers. A main challenge for this industry to overcome is the differences regarding culture and traditions, knowledge and communication, between Information and Communication Technology (ICT) staff and industrial control system staff. Communication is necessary for knowledge transfer, which in turn is necessary to learn from previous incidents in order to improve the incident handling process. This article reports on interviews with representatives from large electricity distribution service operators, and highlights challenges and opportunities for computer security incident handling in the industrial control system space.

Keywords: Information security · Incident response

1 Introduction

In the power and automation industry, there has long been a trend towards more use of Information and Communication Technology (ICT), including Commercial-Off-The-Shelf (COTS) components. At the same time, the threats towards information and the ICT systems that are used to process information are steadily increasing, with Advanced Persistent Threats (APTs) receiving growing attention in the information security community. Organizations face attackers with skills to perform advanced attacks towards their ICT infrastructure, with resources to perform long-term attacks, and with goals of achieving long-term access to the target. In such an environment, organizations must expect that, eventually, their systems will be compromised.

Far from being science fiction, ICT security incidents targeting industrial control systems are already happening. During the last ten years, there have been several examples of power outages or other types of damage to automation and control systems caused by hackers, malicious insiders or software failures. The most famous attack up till now is Stuxnet[1], which appeared during the summer of 2010 as an advanced piece of malware

---

[1] see http://www.symantec.com/connect/blogs/w32stuxnet-dossier.

© Springer International Publishing AG 2016
G. Fahrnberger et al. (Eds.): I4CS 2016, CCIS 648, pp. 129–139, 2016.
DOI: 10.1007/978-3-319-49466-1_9

created to cause physical harm to advanced equipment connected to industrial control systems. However, more recent malware such as Dragonfly, Duqu, Flame and the Night Dragon attack, demonstrate that threats related to reconnaissance and espionage also are relevant for industrial control systems. The underlying threats of these recent attacks are well known to ICT security experts. Such attacks have been around for a long time, and several technical and organizational security measures exist that contribute to reducing the risks. However, there should always be a balance between the accepted level of risk and the amount of investment in security measures. It is impossible to prevent all types of incidents, and thus the ability to appropriately prepare for, and respond to, information security incidents is therefore essential for companies in critical industries that need to ensure and maintain continuous operation of their systems.

This article reports on specific aspects of a larger study on information security incident response management in power distribution service operators (DSOs) [1–3]. Based on a number of semi-structured interviews, we venture to shed light on how communication and cooperation influence how information security incidents are being handled and responded to. We look at responses in terms of both technical measures and human actions, and we pay particular attention to how the follow-up activities are performed; information sharing, lessons learnt, and how experiences in the process control domain are transferred into the overall information security work in the organization. This is studied with respect to both ICT systems and the power automation systems in order to identify possible synergy effects from improved cooperation and communication.

The informants represent three different roles in a set of large DSOs; Chief Information Officer (CIO), Chief Information Security Officer (CISO), and Head of control room/power automation systems. The choice of informants was made based on the intention of identifying current cooperation patterns and possible synergy effects from future cooperation, and viewing the overall management system in general.

2 Background

Incident management is the process of detecting and responding to incidents, including supplementary work such as learning from the incidents, using lessons learnt as input in overall risk assessments, and identifying improvements to the implemented incident management scheme. ISO/IEC 27035 Information security incident management [4] describes the complete incident management process. The process comprises five phases; (1) Plan and prepare, (2) Detection and reporting, (3) Assessment and decision, (4) Responses, and (5) Lessons learnt (see Fig. 1, where the phases have been abbreviated as Plan – Detect – Assess – Respond – Learn). The guideline is quite extensive and would be costly to adopt to the letter, but it is a collection of practical advice, key activities and examples, and is useful for companies establishing their own security incident organization. The ISO/IEC standard addresses corporate systems in general, and does not contain any considerations specifically related to industrial systems. In addition to the ISO/IEC standard, several other guidelines and best practice documents are available.

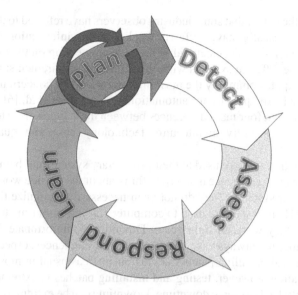

Fig. 1. The Incident Handling Cycle

As indicated by the circular arrow around the Plan phase in Fig. 1, this is where the average organization will spend most of its time, waiting and (hopefully) preparing for the next incident. It is in the planning phase that the groundwork for successful incident management is laid, including establishment of communication channels and an information sharing culture, both between the different disciplines within an organization, and between organizations. Although not the focus of this article, it is also here that other general improvements of the information security controls and mechanisms are performed, guided by evolving best practices and industry standards.

Whereas standards and recommendations exist in the area of incident management, also with respect to critical infrastructures, limited research is available related to managing incidents in an operating environment where automation systems and ICT systems are closely integrated [5]. An efficient incident management process is just as important as technical information security measures when continuous operation is a governing requirement.

3   Stumbling Blocks for Process Control Incident Response

There are usually many obstacles to overcome in order to implement a successful scheme for incident management in an organization. However, the power industry, as well as other process industries, faces one in particular that is not shared by all other industries: the integration of ICT systems and power automation systems; which implies that ICT staff and power automation staff or process control engineers need to cooperate extensively during both daily operations and crisis situations. These two groups differ in several ways, to the extent that some industry observers have referred to them as "zebras and lions". They usually have different backgrounds; information technology or computer science on one side, electrical or process engineering on the other side. They are used to operate different systems with quite different requirements; as an example, confidentiality is quite commonly the most important security concern in ICT systems, while availability is first priority for automation systems. Wei et al. [6] exemplify this further by pointing out four main differences between these types of systems, regarding security objectives, security architecture, technology base and quality-of-service requirements.

Power automation staff are used to their proprietary systems not being connected to any external network[2], and hence not used to think about the outside world as a possible threat towards their systems. They do not even necessarily recognize their systems as actually being ICT. ICT staff are used to computers failing from time to time, needing a re-boot before they work all right again. Downtime is unfortunate, but sometimes necessary, and does not always have large financial consequences, especially not if it is planned. Testing and installing patches is thus business as usual in most ICT systems. In power automation, however, testing and installing patches is extremely difficult, as this most probably leads to some downtime. Downtime can be extremely costly in many industrial control environments, and engineers go to great lengths to avoid it. *If it works, don't touch it* is thus a tacit rule of thumb, which results in large parts of such industrial control systems being outdated and unpatched, and hence vulnerable to a great number of known attacks.

Another difference between ICT personnel and process control engineers is that the former tend to be concerned with *security* (e.g., preventing unauthorized access to information), while the latter are more concerned with *safety* (e.g., preventing a generator from overheating and exploding). Interestingly, *availability* may be an element of both safety and security, but is likely to be interpreted differently by the two camps. The fields of safety and security have different terminologies. As an example, a safety breach may be denoted as a *fault* or an *accident*. Security breaches, on the other hand, may exploit what are denoted as *bugs* or *flaws*[3]. A *safety hazard* may loosely correspond to a *security threat*, but there are substantial differences when it comes to methods and methodologies between the two fields of safety and security.

Recognizing an information security incident is difficult if one is not trained for it. Experiences from the oil and gas industry show that a computer may be unstable for days and weeks without anyone recognizing it as a possible virus infection [7]. Ensuring that the organization detects and handles such an incident is a cultural challenge just as much as a technical one.

Facilitating and achieving understanding and well-functioning collaboration in this intersection between ICT staff and power automation staff will be the most important task on the way to successful information security incident management for process control environments.

---

[2] Although this is generally no longer the case.
[3] A bug is a programming error, while a flaw is a more high-level architecture or design error.

4   Collaboration and Communication in Incident Management

From the literature, we are aware of three main interview studies with the aim of identifying practices regarding incident response. Werlinger et al. [8] studied the practices related to diagnostic works during incident response in a variety of organizations. Ahmad et al. [9] studied incident management in a large financial institution, and Jaatun et al. [7] studied incident response in the petroleum industry. In the following, we summarize the main findings from these studies when it comes to collaboration and communication related to incident management.

Planning of incident management can include a large number of diverse of activities, including getting management commitment, establishing policies, plans and procedures, putting together an incident response team, establishing technical and other support, and performing awareness briefings and training. Studies in the petroleum industry [7] revealed that the organizations usually had several plans covering different aspects of the incident management process, and that there was a need for a short and common plan. It was also found that suppliers were not adequately involved in planning for incidents, although the operator would in many cases depend on them during incident management. Individual information security awareness was also not at a satisfactory level. Scenario training that could have improved this was not performed for ICT incidents as it was done for HSE incidents. Finally, and maybe most disturbing, the study revealed a "deep sense of mistrust" between process control engineers and ICT network administrators.

The identified issues can be interpreted as symptoms of unsatisfactory collaboration and communication when it comes to information security, and incident management in particular. This is disturbing since incident management is collaborative in nature. This is exemplified by Werlinger et al. [8], who found that:

- configuration of monitoring tools for incident response require extensive knowledge of issues that are rarely explicitly documented and obtaining this knowledge may involve also external stakeholders
- the complexity of the IT systems, and also the lack of resources for monitoring, means that incident detection relies on notifications from various stakeholders – including end-users
- verification that there actually is an incident – not a false alarm – may require collaboration with external organizations
- managers often need to be involved in decision making.

The importance of collaboration and communication is also reflected in the procedures for responding to high-impact incidents at the financial organization studied by Ahmad et al. [9]. Technical and business conference calls are set up in order to gather knowledge and communicate progress; in general, the management of the incident relies heavily on communication via teleconferencing, phone, e-mail and the helpdesk system. It is not without reason that Werlinger et al. [8] list 'communication' as one of the five key skills required for diagnosis work.

The challenges related to collaboration and communication are often revealed when studying the learning process that take place after an incident. In the financial institution studied by Ahmad et al. [9] incidents were handled differently depending on their impact ranking. This also influenced the learning process afterwards. For high-impact incidents, the post incident learning process was formalized, and involved at least three meetings. The first two meetings however only included the members of the incident response team. Thus, members from the risk area and the business in general were involved only to a limited extent. For low-impact incidents, the only formal practice was to write a log entry in the incident tracking system. The team involved however still attempted to learn and identify how they could improve based on the experiences from the incident.

The study of the petroleum industry [7] revealed that information security was viewed merely as a technical issue. This technical focus was also found in the study of the financial organization [9]. Especially for low-impact incidents, the emphasis was on technical information, over policy and risk. For high-impact incidents, there was an understanding that it was important to identify root causes that goes beyond the technical issues (e.g. underlying gaps in the processes). However, the learning process also for high-impact incidents involved only technical personnel in the first phases. Reporting from incidents was also technical. Based on the low-impact incidents, several reports were produced for management. This was typically statistical information with a focus on the technical aspects. From the high-impact incidents, the reports were more detailed and a bit broader in scope, but dissemination also to non-technical personnel was not performed satisfactorily. There was a lack of formal policy on how information should be disseminated. In addition, the silo structure of the organization was a hindrance for effective sharing of experiences. The practice can probably be summarized by a finding by Werlinger et al. [8] where the representative from one of the organizations studied explained that security incidents were discussed weekly so that **security practitioners** could learn about new threats and assist in solving challenging incidents.

Despite the obvious weaknesses in the learning process used at the financial organization studied [9], the organization found that the introduction of the formalized learning process for high-impact incidents had resulted in an enormous reduction in the number of high-impact incidents. The importance of learning was also stressed in the study of the petroleum industry [7]. Learning was however considered to be difficult, and the representatives from the industry knew that they did not perform learning at a satisfactory manner. One of the problems highlighted was that they had several reporting systems, of which none was tailored to information security. In addition, the study revealed a lack of openness about incidents. Suppliers were not adequately involved in learning from incidents – although they could play a crucial role. There was also a lack of willingness to share information about incidents to the industry as a whole.

5   Incident Management at Large DSOs - Results from Interview Study

We have performed a large study of information security incident management at several large DSOs [1, 2, 10]. The current introduction of Advanced Metering Infrastructure (AMI) results in – from the point of DSOs – more ICT, and more distributed ICT, and more pathways into their core systems. This has implications for their work on information security, including incident management. In order to know more about the level of maturity in this industry when it comes to management of information security incidents, and also the main challenges faced in this respect, we are performing interviews with key personnel at large DSOs; both personnel from the ICT side and the power automation side. We will in the following highlight the input from the informants that relate to cooperation and communication aspects of incident handling. The results concern practices and experiences in the incident management phases Lessons learnt, Plan and prepare, and Detection and reporting.

5.1 Lessons Learnt

There are large differences between the DSOs when it comes to the post-incident activities. Some report on having routines for regular meetings, or at least evaluation meetings after certain incidents, in order to go through what happened, how they responded, which changes need to be implemented in the near future, and for mutual information exchange. Others do not perform regular reporting or evaluations, neither in the team, nor with the top management. In general, the DSOs do not perform learning activities in connection with incidents with low impact.

Self-experienced incidents form an excellent foundation for internal raising of awareness. Some DSOs seem to do this, whereas others have a rather unstructured approach to these sorts of activities. The respondents state that there are few information security incidents directly related to their industrial control systems, which may influence the experienced level of urgency when it comes to learning activities related to incidents.

The use of indicators or metrics related to information security does not seem to be established. For some specific incidents the interviewees are able to estimate a cost for the work of fixing the problems and reestablishing regular operations. Again, the low frequency of events may contribute to a sentiment that "there is nothing to measure".

5.2 Plan and Prepare

The DSOs do not seem to have established their own Computer Security Incident Response Team (CSIRT) within the organization. They are however required to have an emergency preparedness organization for incidents described in the national regulations for preparedness and contingency. Such incidents are typically those that may have consequences for the power generation and distribution. Some information security incidents can be considered a subset of these.

Plans for incident management are not established in all DSOs. Some are currently working on this, others have not identified the need for such plans. Testing preparedness plans and training on information security incident management routines does not seem to be common practice. This may come as a natural consequence of the lack of plans. Also, this type of exercise does not get high priority compared to other pressing tasks that are needed to ensure daily operations. Some DSOs report that they perform frequent training on emergency preparedness in general, where information security incidents may be part of the problem.

The DSOs confirm that there is insufficient cooperation between ICT and ICS staff, but there are notable exceptions where different respondents from the same company offer dissenting views on this issue. This needs to be studied further, but one possible explanation may be that there are differing views on what constitutes "good coopera-tion". In general, they agree that the evolving Smart Grid landscape with introduce new challenges that will require improved cooperation in the future.

5.3 Detection and Reporting

The respondents all have technical detection systems like IDS/IPS, antivirus, firewalls and similar in place for their administrative ICT systems. For the power automation systems this is a bit more unclear. This lack of clarity may be due to our respondents lacking detailed knowledge; it does not necessarily mean that such detection mecha-nisms are not in place. However, failures, or irregular events, in these systems are just as often detected by the personnel operating them. Also, incidents to the administrative ICT systems are sometimes detected by employees. All respondents report of a culture where reporting is accepted and appreciated, such that the employees are not reluctant to report in the fear of being suspected for doing something wrong.

6 Discussion

Our study of incident management practices and experiences among DSOs reveals much of the same challenges as have been pointed out in other studies. Lack of plans and lack of training was also pointed out as challenges in the petroleum industry [7]. Limited learning activities was also a problem in that industry, and in the study of the financial organization [9]. The reliance on users when it comes to detection of incidents was also reported in the studies by Werlinger et al. [8]. The level of preparedness and the amount of learning activities performed however varies a lot from DSO to DSO. We have seen examples of organizations who are diligent in their application of good incident response practices, but there are also other organizations that are less aware of (or concerned with) incident response in particular, or information security in general.

Our interviews confirm that information security activities often are not considered part of "core business" for the DSOs, but rather more or less implicitly subsumed under the broader category of "ensuring electricity delivery". This also implies that commu-nication between ICT personnel and process control still leaves something to be desired. A naïve cause-and-effect analysis might seem to indicate that the current incident management processes work as intended, and that there thus is no need for improved communication and coordination in this sector. However, we think it is just as likely that the low number of incidents simply means that the would-be attackers at the moment are just having too much fun elsewhere. Just as the malware industry has moved from more or less malicious pranks fueled by idle curiosity to for-profit botnets and extortion, we believe that we have so far seen only the tip of the iceberg when it comes to computer security incidents in the process control industry.

Learning from others' mistakes is generally accepted as being less expensive than learning from your own, and thus there has long been a culture for full-disclosure information sharing[4] among accredited information security incident handling professionals in (particularly academic) Internet Service Providers. The lack of a coordinating Computer Security Incident Response Team in many process control industry segments is also an obstacle to improved information sharing. In Norway, there has been talk of an "Oil CERT" and an "Energy CERT" for years, but the latter was only formally established at the end of 2014[5], and it is still not entirely clear how it will interact with other players. There are commercial alternatives that purportedly document security incidents in the process control industry, such as the RISI database[6], but since these require a quite costly membership to access, it is difficult for independent research groups to assess their content and usefulness.

Despite the interviews giving a clear impression of a practice that seems to work satisfactory, we are hesitant to conclude that the "fire-fighting approach" is something that would work in general. If there are few occurring incidents with high impact, they are likely handled by a single person or a small group, which could imply that tacit knowledge covers who to contact and who does what. It could be argued that introducing a more rigid process with more documentation, reporting, and just "paper work" is not an efficient use of resources if current practice covers the need, but this ignores the perils of relying on the tacit knowledge of a few key individuals. Increasing incident frequency or unplanned absences can require sudden adding of emergency manpower, which will not work well if there is no documentation.

7   Conclusion

The interviews reveal a lack of systematic approaches to information security incident management. They also show that there is not a close cooperation between ICT staff and power automation staff, but rather quite clear definitions of responsibilities, although some respondents from the same DSO have given quite opposite opinions on this matter.

Even though the interviews portray a current practice that seems to work in a satisfactory manner, we advise against relying on this for the future, also because most of the respondents see future challenges for information security incident management in smart grid (and, by extension, industrial control system) environments. They confirm our initial concerns regarding the need for much closer cooperation between ICT staff and power automation staff, and the challenges related to this cooperation.

8   Further Work

In general, there is a need for additional empirical research in the area of incident response management [5, 11]. Although we get the impression that the informants

---

[4] http://www.first.org/.
[5] https://www.kraftcert.no/.
[6] http://www.securityincidents.org/.

describe current practice, not only an ideal picture, we would like to do some further investigations on this matter. One way to approach this would be to run a retrospective group interview after a DSO has experienced an ICT security incident, i.e., a meeting with the persons and/or parties involved in solving the incident, where the complete course of events would be analyzed in order to understand how the organization responded to that specific incident. Interesting aspects for further exploration include how the incident was detected, reported and resolved, in which ways they followed their plans, and if not, how and why the plans were abandoned, and so on. This could be done as part of the organization's own evaluation process, if such exists.

There are usually quite a few differences between theory and practice. Observations are therefore also important. While the interviews give much insight in how incident management is planned and performed, observing the work in practice will give invaluable additional knowledge. Having knowledge of both theory and practice will make it possible to compare theory and practice, suggest realistic improvements, and hence actually make a contribution to the industry. Ideally, researchers should be present during a certain period of time in locations where incidents are detected and responded to. Observing meetings and other interactions, and having informal talks with the involved personnel by the coffee machine can also be important sources of information. The intersection between ICT and power automation competence, culture, language, incidents, and tools will be especially interesting to observe.

Regular emergency preparedness exercises have long been required in many critical infrastructure sectors, but these have until recently not considered cyber security a necessary component. There is a need to perform more empirical research on how preparedness exercises can incorporate cyber security, and to evaluate how this contributes to better cyber security for an organization [11].

Acknowledgments. The authors would like to thank the distribution system operators who have contributed with informants for our interviews. This research has been supported by the Norwegian Research Council through the projects DeVID and Flexnett.

References

1. Line, M.B.: A case study: preparing for the smart grids - identifying current practice for information security incident management in the power industry. In: 2013 Seventh International Conference on IT Security Incident Management and IT Forensics (IMF), pp. 26–32 (2013)
2. Line, M.B., Tøndel, I.A., Jaatun, M.G.: Information security incident management: planning for failure. In: Proceedings of the 2014 Eighth International Conference on IT Security Incident Management and IT Forensics, pp. 47–61. IEEE Computer Society (2014)
3. Line, M.B., Tøndel, I.A., Jaatun, M.G.: Current practices and challenges in industrial control organizations regarding information security incident management - does size matter? Information security incident management in large and small industrial control organizations. Int. J. Crit. Infrastruct. Prot. **12**, 12–26 (2016)
4. ISO/IEC 27035:2011 Information technology - Security techniques - Information security incident management. ISO/IEC (2011)

5. Tøndel, I.A., Line, M.B., Jaatun, M.G.: Information security incident management: current practice as reported in the literature. Comput. Secur. **45**, 42–57 (2014)
6. Wei, D., Lu, Y., Jafari, M., Skare, P.M., Rohde, K.: Protecting smart grid automation systems against cyberattacks. IEEE Trans. Smart Grid **2**, 782–795 (2011)
7. Jaatun, M.G., Albrechtsen, E., Line, M.B., Tøndel, I.A., Longva, O.H.: A framework for incident response management in the petroleum industry. Int. J. Crit. Infrastruct. Prot. **2**, 26–37 (2009)
8. Werlinger, R., Muldner, K., Hawkey, K., Beznosov, K.: Preparation, detection, and analysis: the diagnostic work of IT security incident response. Inf. Manag. Comput. Secur. **18**, 26–42 (2010)
9. Ahmad, A., Hadgkiss, J., Ruighaver, A.B.: Incident response teams – Challenges in supporting the organisational security function. Comput. Secur. **31**, 643–652 (2012)
10. Line, M.B.: Understanding information security incident management practices: a case study in the electric power industry. Ph.D. Thesis, NTNU (2015)
11. Bartnes, M., Moe, N.B., Heegaard, P.E.: The future of information security incident management training: a case study of electrical power companies, Computers and Security (2016)

Routing and Technology

Routing over VANET in Urban Environments

Boubakeur Moussaoui[1], Salah Merniz[2], Hacène Fouchal[3],
and Marwane Ayaida[3(✉)]

[1] Laboratoire D'Electronique et des Télécommunications Avancées (ETA),
Université Mohamed Bachir El Ibrahimi de Bordj Bou Arreridj,
34031 Bordj Bou Arréridj, Algeria
Moussaoui.bkr@gmail.com
[2] Université de Constantine-Abdelhamid Mehri, Constantine, Algeria
s_merniz@hotmail.com
[3] Centre de Recherche CReSTIC, Université de Reims Champagne-Ardenne,
51687 Reims Cedex 2, France
{hacene.fouchal,marwane.ayaida}@univ-reims.fr

Abstract. Experimental deployment of Cooperative Intelligent Transport Systems have been undertaken these last years. But a real deployment is lower than expected. One of the main reasons is the high cost
of investments of Road Side Units on roads. Road operators need a lot
of money in order to achieve this deployment. We suggest to reduce
this investment by the deployment In this paper, we propose a combination of GPSR (Greedy Perimeter Stateless Routing) and an extension
of Reactive Location Service denoted eRLS. They used to be combined,
i.e. GPRS takes care of routing packets from a source to a destination
and eRLS is called to get the destination position when the target node
position is unknown or is not fresh enough. When a destination is not in
the area of the sender, the exact position of the target is first looked for.
An extra overhead is generated from the sender to the receiver since t is
quite complicated to have an efficient Location System service as RLS.
In the meantime, in deployed Cooperative Intelligent Transport Systems
(C-ITS), fixed components are usually installed and denoted road side
units (RSUs). In this paper we suggest to use these fixed RSUs to achieve
the location service.

When a sender needs to send a packet, it first looks for the next RSU
and sends the packet to it This RSU is connected to all other RSUs using
a wired network. The aim is to find the nearest RSU to the destination.
It is done thanks to a request on all RSUs Then the packet is forwarded
to it. The last step is to look for the target node in the area of this last
RSU. Even if the node has moved, we will have more chances to reach
the node in this area.

Experimentations of our proposes solution have been done over the
NS-3 simulator where an extension of RLS is developed and included on
the simulator. The obtained results have shown a better accuracy vehicle
locations and better performances in terms of delay.

Keywords: VANETs · Location-based services · Geographic routing
protocols

© Springer International Publishing AG 2016
G. Fahrnberger et al. (Eds.): I4CS 2016, CCIS 648, pp. 143–152, 2016.
DOI: 10.1007/978-3-319-49466-1_10

1   Introduction

VANETs (Vehicular Ad-hoc NETworks) are studied by many teams since amore than a decade, they are the most practical case of MANETs (Mobile Ad-hoc NETworks). Topology-based routing protocols are recognized as less efficient in such networks due to the high network dynamics. Geographic routing protocols are used since they offer better results for such networks. Each node knows its actual geographic position and the target node position as well. We badly need a location-based service in order to get the geographical position of the destination. The practical use of VANET is C-ITS (Cooperative-Intelligent Transport Systems) where Road Side Units (RSU) are deployed along roads. These units are not deployed as massively as we could expect, they are quite expensive. They are considered as fixed nodes in a VANET. They built somehow the network infrastructure.

In this paper, we intend to use these units to built a location-based service. In fact these units are all connected to the internet with a wired connections. They are also connected to a main server able to maintain nodes positions. When a node s needs to reach a destination d, it has to look for its position. The node s needs to send a packet, we first look for the next RSU and send to it the packet. This RSU is connected to all other RSU using a wired network. The aim is to find the nearest RSU to the destination d. Then the packet is forwarded to it. The last step is to look for the target node in the area of this last RSU. Even if the node has moved, we will have more chances to reach the node d in this area.

For this purpose, we have proposed a patch over the NS-3 simulator which mixes both GPSR and an extended version of RLS (eRLS) according to our proposal. We have undertaken experimentations and we have considered the freshness of the destination location data. On one hand, we have proved that the proposed scheme provides promising results in terms of latency, packet delivery rate and overhead and on the other hand the freshness of the location data has a real impact on the network performances. We have also proposed a solution de to deploy a location service.

The paper is organized as follows. Section 2 is dedicated to related works. Section 3.3 details our combination algorithm about GPSR and HLS. Section 4 explains experimentations and the obtained results. Finally, Sect. 5 gives a conclusion and some hints about future works.

2   Related Works

In this section, we will describe some related works, published in this last decade. All these works focused on routing problem in VANETs, and response to the targeted problem using V2V communication and/or V2I communication with more or less effective manner. They do not consider the changed position during the dissemination. Both the source node and destination are vehicles in general case. To deal with to the high moving speed and dynamic changes of topology, Road Side Units, can provide a lot of benefits to the routing task. In order to get more realistic solution, it is not necessary to deploy at each intersection an RSU. For more details about routing among mobile and vehicular ad hoc networks, the reader may refer to [2].

In [3], authors propose an improved version of DGRP (Directional Greedy Routing Protocol). In addition to the information used by DGPR (position, speed, direction) provided by GPS, they consider and evaluate link stability. As result, they reduce links breakage, enhance reliability of a used path and ensure a high packet delivery ratio.

Another proposed work which assist routing in VANETs using base station was presented in [4]. Roads are organized in segments, each of them is governed by a base station, this end response to a route request packet by choosing the shortest path using IVC, otherwise (no available path) it (RSU) checks if there is enough free bandwidth to grant the request. They employed two mechanism, using Fast Learning Neural Networks, to prevent link failure and predict bandwidth consumption during handoffs. An alternative path will be established, before that a broken link occur.

Authors in [5] have proposed an infrastructure-assisted routing protocol. The main benefits behind this approach, was to reduce the routing overhead and improve the end-to-end performance. A backbone network connects using Road Side Units, an enhancing search of the shortest path to reach the destination is made by this infrastructure. Authors present an extended protocol to the known topology-aware GRS routing protocol. At the intersection, Anchor nodes in GRS compute the shortest path to reach the destination using Djikstra algorithm without taking any consideration to the density of roads or any other parameter which can ensure connectivity aware these roads. Roadside-Aided Routing (RAR) was proposed in [6].

2.1 Location-Based Services

Location-based services can be classified into two classes: "Flooding-based" and "Rendez-vous-based". The first class is composed of reactive and proactive services. In the proactive flooding-based location-based service, every node floods its geographic information through all the network periodically. Thus, all the nodes are able to update their location tables. Since this approach uses flooding and may surcharge the network by location update messages, several techniques to reduce the congestion were used. One of them is to tune the update frequency with the node mobility (the more node is moving fast, the higher update location frequency is used).

Therefore, the update frequency decreases with the distance to the node. The second idea is, a node with high mobility sends more update location packets. As a result, there are less packets than a simple flooding scheme without affecting the network performances. For the second group (i.e. the reactive flooding-based location-based service), the location response is sent when receiving a location request. This avoids the overhead of useless location information of some nodes updated and never used. But, it adds high latencies not suitable in VANETs. One of these known services is Reactive Location Service (RLS) [7].

In the second class (rendez-vous-based location service), all the nodes agree on a unique mapping of a node to other specific nodes. The geographic information are disseminated through the elected nodes called the "location servers".

Thus, the location-based services consists of two components:

1. Location Update: A node has to recruit location servers (chosen from other nodes) and needs to update its location through theses servers. The location servers are responsible of storing the geographic data of the relating nodes.
2. Location Request: When a node needs to know the location of another node, it broadcasts a location request. The location server will replay as soon as it receives this request.

2.2 Geographic Routing Protocols

Routing protocols algorithms must choose some criteria to make routing decisions, for instance the number of hops, latency, transmission power, bandwidth, etc. The topology-based routing protocols suffer from heavy discovery and maintenance phases, lack of scalability and high mobility effects (short links). Although, geographic routing are suitable for large scale dynamic networks. The first routing protocol using the geographic information is the *Location-Aided Routing (LAR)* [8]. This protocol used the geographic information in the route discovery. This latter is initiated in a *Request Zone*. If the request doesn't succeed, it initiates another request with a larger *Request Zone* and the decision is made on a routing table. The first real geographic routing protocol is the *Greedy Perimeter Stateless Routing (GPSR)* [1]. It is a reactive protocol which forwards the packet to the target's nearest neighbor (Greedy Forwarding approach) until reaching the destination. Therefore, it scales better than the topology-based protocols, but it does still not consider the urban streets topology and the existence of obstacles to radio transmissions. Another geographic routing protocol is the *Geographic Source Routing (GSR)* [9]. It combines geographical information and urban topology (street awareness). The sender calculates the shorter path (using Djikstra algorithm) to the destination from a map location information. Then, it selects a sequence of intersections (anchor-based) by which the data packet has to travel, thus forming the shortest path routing. To send messages from one intersection to another, it uses the greedy forwarding approach. The choice of intersections is fixed and does not consider the spatial and temporal traffic variations. Therefore, it increases the risk of choosing streets where the connectivity is not guaranteed and losing packets.

In [10], authors propose an improved version of DGRP (Directional Greedy Routing Protocol). In addition to the information used by DGPR (position, speed, direction) provided by GPS, they consider and evaluate link stability. As a result, they reduce links breakage, enhance reliability of a used path and ensure a high packet delivery ratio.

3   Adapted Location-Based Service

3.1   eRLS General Description

In order to understand our proposed location service *eRLS*, we introduce the
Fig. 1. This figure represents an example of a communication scenario in an
urban scenario. We assume that the source vehicle *V0* needs to send data to
the vehicle *V10*.

In the background, when a node passes close to a Road Side Unit (RSU), the
node sends a *Hello* messages in order to inform the RSU of its position. Each
time slot, every RSU sends last vehicles' positions to the **Location Server**.
The latter stores these locations after processing it. Then, it waits for a location
request from a RSU and will replies with the new position of the vehicle if it can
find it.

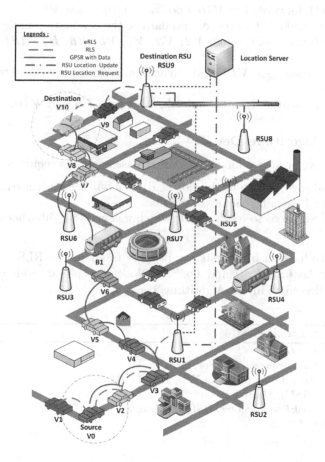

Fig. 1. An example of a scenario with eRLS

The communication between **V0** and **V10** could be divided into some steps:

1. The source node **V0** starts to broadcast a request for the **V10**'s position using **eRLS** through its neighbors (**V1** and **V2**) to find the nearest RSU.
2. The nearest RSU, the **RSU1** receives the request through the nodes **V2** and **V3**.
3. The **RSU1** forwards this request to the **Location Server**.
4. The **Location Server** searches in its databases for the last RSU that notifies the **V10**'s presence (**RSU9**).
5. The **Location Server** sends then a location request for the node **V10** to the **RSU9**.
6. The location request reaches **V10** through the node **V9** using **RLS**.
7. The node **V10** replies to **RSU9** through the node **V9** with its new position.
8. The new **V10**'s position is then sent to the **RSU1** via the **Location Server**.
9. The **RSU1** forwards the **V10**'s position to the node **V0**.
10. The source node **V0** starts to send data to the destination node **V10** using the **GPSR** protocol through **V2**, **V3**, **V4**, **V5**, **V6**, **B1**, **RSU6**, **V7** and **V9**.
11. The destination node **V10** sends an acknowledgment backward to the source node **V0**.
12. The communication is now established between **V0** and **V10**.

3.2   eRLS Algorithms Details

In our study, we need two kinds of location based-services requests:

- When a node needs to know the closest RSU, it will send a recursive broadcast requests until reaching the next RSU.
- When an RSU needs to reach a node which was in its neighborhood some time earlier.

In order to have such requests, we propose to adapt the RLS protocol with two requests: *LookForRSU* and *LookForNode*. The pseudo-code of these two requests are described in the Algorithms 1 and 2.

Algorithm 1. LookForRSU

1: *get NeighbourList*
2: *send broadcast request LookForRSU*
3: **while**(*not found and* $j < Size(NeighbourList)$)
4: **if** ($N_i == RSU$) **then**
5:     $Adr = LookForNodeOverRSU$ (*destination*)
6:     $found = TRUE$
7: **else**
8:     $i + +$
9: **end if**
10: **end while**
11: **return (adresse)**

Algorithm 2. LookForNode

1: *get NeighbourList*
2: *send broadcast request LookForNode*
3: **if** *(found == TRUE)* **then**
4:     send(Result of *RLS(destination)*) to source
5: **else**
6:     continue broadcasting
7: **end if**

In order to use the location based-device, we should maintain location information over RSU and a main location server. In order to do so, we run the *maintainLocations* continuously.

3.3   Combining Routing with eRLS

The combination of the *eRLS* with *GPSR* is described in the Algorithm 3.

Algorithm 3. Combined eRLS-GPSR

1: Intialization: Source s, Destination d, Data D
2: RSUd <- eRLS(d)
3: **if** d is in neighborhood of RSUd **then**
4:     nextHop <- RSUd
5:     GPSR(s, d, nextHop, D)
6: **else**
7:     nextHop <- Closest RSU to s
8:     GPSR(s, d, nextHop, D)
9:     Send D to RSUd from nextHop using RSU backbone
10:     **wihle** (d not reached)
11:         nextHop <- RLS(d)
12:         GPSR(s, d, nextHop, D)
13:     **end while**
14: **end if**
15: GPSR(d,s, $nextHop^{-1}$, Ackdata)
16: Location update
17: **wihle** $(TRUE)$
18:     update RSU tables
19:     In each slot get all location tables and gather all of them
20:     send($RLS(destination)$) to source s
21: **end while**

First, if a node s has a data D to send to the destination d, it used the algorithm *eRLS* presented in the Algorithm 1 to obtain the *RSUd*, which is the closest RSU to the destination. If s is in the same area than *RSUd*, s forward the data packet to *RSUd*. Otherwise, it sends the data packet to the closet RSU.

The latter transmit using wired connection the data packet to *RSUd*. Knowing that d passed close to *RSUd* recently, it sends a *RLS* request to know how to get to d. After that, the data packet is transmitted in multi-hop until d. Then, d will send a location update to s if it has more data to send in order to use directly the V2V links.

In the background, all RSUs collect the vehicles' positions when they pass through. Moreover, a server collects all the neighbors location tables from RSUs. It also replies to the *eRLS* requests with the closest RSU when another RSU asks for a node.

4   Simulations

4.1   Working Environment

The simulations were performed using the Ns-3 simulator 2.33 [11]. The geographic routing protocol used is the Greedy Perimeter Stateless Routing (GPSR). The chosen area is a 2×2 km^2 of a real map representing part of the French city *Reims*. This area is extracted from Open Street Map [12]. The MAC layer used is 802.11p [13]. The parameters used in the simulation are summarized in the Table 1.

Table 1. The simulation parameters

Parameters	Value
Channel type	Channel/WirelessChannel
Propagation model	Propagation/TwoRayGround
Network interface	Phy/WirelessPhyExt
MAC layer	802.11p [13]
Interface queue type	Queue/DropTail/PriQueue
Link layer	LL
Antenna model	Antenna/OmniAntenna
Interface queue length	512 packets
Ad-hoc routing protocol	GPSR
Location-based service	HLS/HHLS
Location cache maximum age	4, 8, 12, 16, and 22 s
Area	2×2 km^2
Number of nodes	50
Simulation time	150 s
GPSR beacon interval	0, 5 s
CBR traffic	4×100 packets/node
CBR packet size	128 KB
CBR sent interval	1 s

At each simulation, every node initiates 4 CBR traffics of 100 packets with a size of 128 KB to 4 random destination nodes with a second of interval between each sent. The CBR traffic simulates for example an audio or a video streaming. It may be used in security applications, such as viewing the video stream from a camera located on a bus by the police car or the security agent vehicle. Also, this traffic could be used in entertainment applications to connect to the Internet or to play online video games.

4.2 Experimentation Results

Our experimentations have provided the following figures. On each figure, we show the network behavior with usual RLS (denoted RLS curve) and the behavior of our proposed solution (denoted eRLS curve for enhanced RLS). The following Fig. 2 shows the delay to send a message depending on the size of the network from 10 nodes to 100 nodes. we observe that the delay to send a message from a node s to d decreases with eRLS even if the number of nodes is higher. The gain is around 10 %.

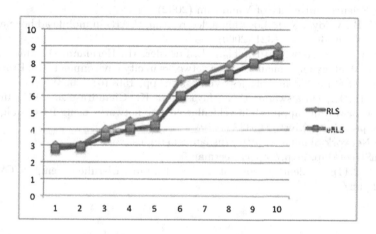

Fig. 2. The mean delay to send a packet

5   Conclusion and Future Works

We have proposed an extension of the RLS service using fixed nodes (RSUs). We have proposed a complete algorithm able to transmit a packet from a mobile node to another one traversing a network composed of mobile nodes and fixed nodes. The fixed nodes are the core system of our location service. They have been very useful to ensure the localization of nodes. The delay for a packet to traverse the network is decreased with eRLS. We have implemented over NS-3 a path of extended RLS (eRLS) which have shown a very interesting enhancements in terms of delay, lost packets.

References

1. Karp, B., Kung, H.T.: GPSR: Greedy Perimeter Stateless Routing for wireless networks. In: 6th International Conference on Mobile Computing and Networking - MobiCom00, New York, USA, pp. 243–254 (2000)
2. Rana, R., Rana, S., Purohit, K.C.: A review of various routing protocols in VANET. Int. J. Comput. Appl. **96**(18), 28–35 (2014)
3. Prasanth, K., Duraiswamy, D.K., Jayasudha, K., Chandrasekar, D.C.: Improved packet forwarding approach in vehicular ad hoc networks using RDGR algorithm. arXiv preprint arXiv:1003.5437 (2010)
4. Huang, C.J., Chuang, Y.T., Chen, Y.J., Yang, D.X., Chen, I.F.: QOS-aware roadside base station assisted routing in vehicular networks. Eng. Appl. Artif. Intell. **22**(8), 1292–1301 (2009)
5. Borsetti, G.J.: Infrastructure-assisted geo-routing for cooperative vehicular networks. In: IEEE Vehicular Networking Conference (VNC), pp. 255–262 (2010)
6. Peng, Y., Abichar, Z., Chang, J.M.: Roadside-aided routing (RAR) in vehicular networks. In: IEEE International Conference on ICC 2006
7. Kasemann, M., Hartenstein, H., Mauve, M.: A reactive location service for mobile ad hoc networks. Technical report, TR02014, pp. 121–133, Department of Computer Science University of Mannheim (2002)
8. Ko, Y.-B., Vaidya, N.H.: Location-aided routing (LAR) in mobile ad hoc networks. Wirel. Netw. **6**(4), 307–321 (2000)
9. Lochert, C., Hartenstein, H., Tian, J., Fuessler, H., Hermann, D., Mauve, M.: A routing strategy for vehicular ad hoc networks in city environments. In: Proceedings of the IEEE Intelligent Vehicles Symposium, pp. 156–161 (2003)
10. Prasanth, K., Duraiswamy, D.K., Jayasudha, K., Chandrasekar, D.C.: Improved packet forwarding approach in vehicular ad hoc networks using RDGR algorithm. arXiv preprint arXiv:1003.5437 (2010)
11. NS2: Network Simulator. http://nsnam.isi.edu/nsnam/
12. OpenStreetMap. http://openstreetmap.fr/
13. NS2 802.11p Implementation. http://dsn.tm.uni-karlsruhe.de/english/Overhaul_NS-2.php/

Tech4SocialChange: Technology for All

André Reis[1]([✉]), David Nunes[1], Hugo Aguiar[1], Hugo Dias[1],
Ricardo Barbosa[1], Ashley Figueira[1],
André Rodrigues[1], Soraya Sinche[1], Duarte Raposo[1], Vasco Pereira[1], Jorge Sá Silva[1],
Fernando Boavida[1], Carlos Herrera[2], and Carlos Egas[2]

[1] Department of Informatics Engineering, University of Coimbra, Coimbra, Portugal
afreis@student.dei.uc.pt
[2] Escuela Politecnica Nacional, Quito, Pichincha, Ecuador

Abstract. Universities and other educational institutions are sometimes accused of not being involved in real world problems, focusing more on the scientifically value of the work produced and not on the humanitarian value. A way of encapsulating the second with the first is the main goal of the Tech4SocialChange that is described here. An innovative database/repository of challenges with real impact in the world is created and given access to people with skills and knowledge to tackle them. Also the work made by researches can be stored and used in a project and the researcher gets recognition for it by becoming referenced in that project. A web application has been built as a prototype for this process and can be accessed in www.tech4socialchange.org. It has been planned and developed by a team of students and researchers of the Department of Informatics Engineering of the University of Coimbra and is currently being constantly altered according to feedback received by the testers in the same team. This paper presents an application that aims to help people that face certain challenges every day and motivate those that have the skillset, to tackle these challenges, into doing so.

Keywords: Social problems · University-society relation · Innovation · Entrepreneurs · Problem solving

1 Introduction

By taking advantage of the academic world, which is sometimes accused of not taking into account real world situations, and providing a database/repository of problems that have a real and direct impact in the lives of people someplace in the world, the means and/or knowledge to tackle these problems are provided.

Tech4SocialChange's goal is to be this bridge that links universities to the problems affecting people around the world.

Researchers that sometimes struggle to find interesting subjects for their work get a database where they can consult and start building solutions to be applied in the real world. After publishing their work, they can also share it with the community, so it can be used in projects that have an impact.

© Springer International Publishing AG 2016
G. Fahrnberger et al. (Eds.): I4CS 2016, CCIS 648, pp. 153–169, 2016.
DOI: 10.1007/978-3-319-49466-1_11

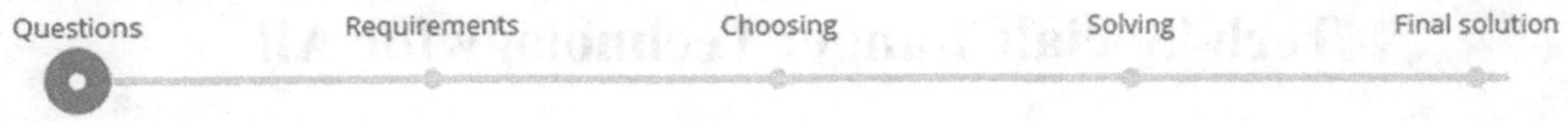

Questions   Requirements   Choosing   Solving   Final solution

Fig. 1. Phases of a problem

Students are also an important part of this process. Typically, they have assignments in their courses that have a purely academic value. By letting students work on real world problems, the assignments would gain an increased value with real impact in the world, which also gives the student an increased value and knowledge for their professional career.

The problems can be submitted by anyone that faces or has knowledge of some kind of challenge or difficulty, in either their own or someone else's life, and that would like to see it solved/tackled by people with the skills to do so.

The process that leads to the solving of a problem is incremental. First it needs to be clear and well defined to let the solvers understand the context and needs of the problem. Next, based on the information provided about it, the solvers need to come up with ideas and develop a project that answers all or most of the problem's needs.

After this, a project must be chosen to be developed and applied in the real world. This choice must be made by the ones that are closer to the problem. In this case, they are the ones who submitted it in the first place. However, some people might not have the skills or knowledge to verify if a project actually responds to all of the requirements in the problem's description. For this situation, specialists are needed; they are people with experience in analyzing and verifying the requirements of a project and validating these in respect to the problem.

In this process, three types of users have been identified:

- Problem Makers create/submit problems.
- Problem Solvers come up with ideas and create projects with the intent of solving the problems.
- Problem Specialists have experience in matching a problem's needs and a project's requirements.

Also, all three types of users can help in the first phase (definition of a problem) by asking questions about the problem to the Problem Makers. In the second phase, the Problem Solvers use the information gathered to think up solutions and create projects. In the third phase, with the help of Problem Specialists that recommend projects to the Problem Makers, a decision has to be made about which of the proposed projects will be developed in the next phase to, ultimately, solve the problem.

The fourth phase is the actual development of the project. In this phase, a simple tracking method is provided with a task list. This list is updated every time a new step in the project planning is created or concluded. This way, the Problem Solvers can give feedback on the project's progress.

The fifth and final phase corresponds to the project in its final state: ready to be implemented and used in the real world. It is the responsibility of the project's team to evaluate if its final state has been reached.

The five phases can be better observed in Fig. 1 where each one is named respectively.

To submit, and to help with the definition of the problem, some questions are provided in the beginning to the Problem Makers. These questions act as guidelines and must be answered before presenting the problem to the community:

- What problem do you want to solve? Or what do you want to change?
- Why does this problem exist?
- What is going to change in the world after the problem is solved?
- What product could be invented? What impact it should have?
- Is there a complete or partial solution to this problem? What is its limitation?
- Do you have something that might support a solution?

The first two guidelines aim to gather information about the environment in which the problem occurs. The next two are about what to expect of the solution to build and the impact that it will cause in response to the problem. The last two guidelines are optional and refer to the existing alternatives or solutions to the problem: why they don't apply to this problem in specific and if there are resources available that can be used in the final solution.

2   State of the Art

There are already projects with the goal of tackling social problems and developing solutions to them. Tech4SocialChange innovates over these through a novel process of processing problems, which is described in Sect. 3.

2.1   HeroX

HeroX [1] is a profitable platform that allows anyone to create a competition and define the conditions for its completion. These competitions are funded by whoever launches them and are based on unsolved problems that are to be solved by combining crowd-sourcing, competition and collaboration.

A challenge is an online competition where people all over the world have the opportunity to solve or build a solution. The winner gets prize money, awarded by the entity that created the competition. To help turn a problem into a challenge, some guidelines [2] are provided by HeroX:

- What problem do you want solved?
- Why does this problem exist?
- What breakthrough are you committed to creating?
- What is the "finish line" or bullseye?
- How long will this challenge last from day 1 to day "won"?

These competitions are managed by an HeroX team that takes care of team selection (they choose who enters the competition or not; the criteria for this choosing depends on the requirements initially set for the competition), management and judging.

To participate in a competition, the users must pay a fee; this ensures that the competitors are committed to finding a solution to the problem being tackled. It also creates a sense of assurance to the entities or groups funding these competitions.

A competition's winning conditions are set on the beginning and are used by the HeroX team to determine the end victor.

The rights to the final solution can differ from challenge to challenge. They can be attributed to the developers, the creators of the competition, HeroX or put under some specific license.

2.2   OpenIDEO

It is a platform that allows for the splitting of the innovation process into phases and building on the ideas of people.

Challenges and programs are created, using crowdsourcing, as a means to tackle problems around the world.

A challenge can last from three to five months and is focused on a single issue that the community can work on and find and develop a solution. A program is a long-term partnership where a specific grand issue (climate change, for example) is tackled and numerous challenges, events or other activities are released [3].

All challenges require financial sponsorship to cover their own costs associated with managing and providing tech and community support. This approach is based on IDEOs design thinking. Tim Brown, CEO of IDEO, states that:

"Design thinking is a human-centered approach to innovation that draws from the designer's toolkit to integrate the needs of people, the possibilities of technology, and the requirements for business success" [4].

This methodology uses skills people have but get overlooked by the standard/popular problem-solving methods. This concept allows for the final solution to be emotionally meaningful and also functional as it integrates feeling, intuition and inspiration with rational and analytical. There are three concepts to keep in mind: inspiration (is the problem or opportunity that motivates the search for solutions), ideation (process of generating, developing and testing ideas) and implementation (path that leads the project to the real world).

In the end, there is a selection of winners, chosen by the sponsor of the challenge and the OpenIDEO team. This selection is based on the challenge criteria and on the OpenIDEO team's skills to implement it. All other ideas can be developed further and used/adapted on other challenges that meet their purpose.

2.3   Others

A. Innocentive

Innocentive uses crowdsourcing solutions that are built for business, social, policy, scientific and technical challenges. These challenges are competitions where the objective is to find a solution to a problem that a client (group or company) has submitted into Innocentive. This submission is based on some criteria, relevant information about the problem (Innocentive helps determining what is important or not) [5]. Also, the winner is determined by the entity that created the challenge and also the award. Innocentive can help with the winner selection but ultimately the decision is of who submitted the problem.

After a challenge ends, the whole solution is given to the entity, including the rights.

The problem solving network and tools that Innocentive already has presents a big motivation for groups, companies and other entities to submit their problems and have them being solved by other people. To the solvers (single individuals or teams), the prizes that are awarded are the main motivation to use Innocentive to work on the solving of problems presented in the platform.

B. CodeForAmerica

CodeForAmerica partners with local governments to build open-source technology and train groups of people to improve government services. It focuses on four key government services:

- Health and human services
- Economic development
- Safety and justice
- Communication and engagement

The way CodeForAmerica gets people and governments to participate is through a fellowship program. This program joins technologists and local governments across America for a year, while working full-time. During this period, the technologists become a part of the community, researching user needs, meeting with stakeholders and proposing solutions. This way, with collaboration from the government, it is possible to build technology that is user-centered and data-driven [6].

The final product of this fellowship is, generally, an early stage application that improves a government service or function. The period of fellowship is a way to encourage innovation and improve risk tolerance inside the government.

Every year, eight to ten governments are selected and twenty-four to thirty people are chosen to the fellowship program through a competitive selection. The government selected has to be in the United States and has to want to work on projects involving health, economic development, safety and justice. They also have to provide support to the technologists that are helping and also be able to support the work that these leave by the end of the fellowship.

To enlist in the fellowship as a fellow/technologist, an application has to be submitted through the page in the website of CodeForAmerica.

Usually a fellowship costs 440.000 dollars. Of this investment, 50 % is covered by the local government, to cover for expenses of the team (benefits, travel, training, salary). The other half of the investment is raised by the government with the help of corporations, foundations and individuals, which helps cover the costs of management of the fellowship.

C. Hack4Good

Hack4Good [7] is an event where any technology enthusiast can participate.

Each event has problem as its main theme. The goal is to find a way to solve or change people's actions towards this problem. The event is global; in a single day, groups of people around the world gather to build prototypes that address different challenges inside the problem.

The problem is divided into challenges to let people focus on more specific issues instead of trying to find a solution to a broader one, like climate change. These challenges and the problem itself are set by NGOs, government organizations and experts in fields related to the problem.

Teams of solvers have one days to find a solution, create a prototype that tackles the problem and make a deep impact in the world. Judges, in each location, are made up of technological experts and are from fields related to the problem being addressed at the event.

The judging is based on the potential of impact that a solution might cause. After a first selection, a solution will move on to a judging at the global scale, competing with the best selected from other locations with different judges.

There is no specific prize for the winners. At 12 of September of 2014, where the theme was climate change, the solutions selected at the global scale were presented as a part of New York City Climate Change Event alongside with the United Nations Climate summit.

3   Tech4SocialChange

Students often have assignments with mere academic value. If these assignments could be directed to real world problems then not only it would serve as an increase in the assignment's value, but also as a real world work experience.

Teachers sometimes struggle to find exciting work subjects that incorporate all of the essential class material. Having a place where they could find projects or subjects that allow the students to come up with ideas to work on during the course would be a major help. This also works for researchers that have difficulties finding exciting subjects to apply their work, or even share their results with people involved in projects with a real or big impact in the world.

As such, all these people would be making the world a better place while improving the value of their own work. This is Tech4SocialChange's audience and goal: to bring the academics' technical knowledge closer to solving the world's problems.

In this section we present a prototype (1) that implements the process presented in Sect. 1. This prototype consists on a web application that has the following main objectives:

- Create a user
- Create problems
- Create projects associated with problems
- Correctly manage the phases in which a problem is presently in
- Allow researchers to submit their work
- Reference researches in projects

3.1 User Perspective

There are three roles already mentioned before: Problem Maker, Problem Solver and Specialist. In addition to these, there is the researcher. This last one is not a role but a type of user, as any user can submit research and thus be recognized as a researcher in the application.

A. Create User

To register a user, it is requested the input of the name of the user, an email to use in the login and a password.

After the registration, an email is sent to the address provided, asking the user to verify its account. This helps in identifying the active users in the application. If, after two days, the user doesn't activate his account, it is deleted.

When the activation is done, the user can login into the application with the email and passwords provided.

The first time the user logs in, it is prompted to choose whether it wants to become a Problem Maker or a Problem Solver, and if he wants to become a candidate to Problem Specialist. This choice can be viewed in Fig. 2.

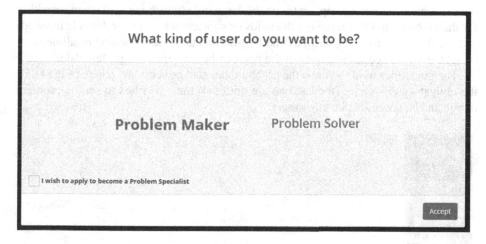

Fig. 2. Choosing the type of user

B. Create Problem

If the user chooses to become a Problem Maker, he can create a problem in the platform. To create the problem, only a title is asked initially. Problems have three states:

- Public, every user in the application has access to the problem
- Private, only the Problem Makers in the problem's team and the Problem Solvers in the solution's team can see the problem
- Draft, only the Problem Makers in the problem's team can see it

After submitting the title of the problem, its state is set to private and its phase to the first, with no deadline set.

Fig. 3. Editing a problem

The problem can be edited in the page shown in Fig. 3. This page asks for the title of the problem, a brief description (or pitch), keywords (which are used when searching for the problem and that represent the fields or subjects where the problem is inserted) and the deadlines for the Questions and Requirements phases. The first deadline must be set before the second and none of the two can be set before the present date.

The guidelines used to define the problem can also be set in the page of Fig. 4. All the obligatory guidelines (the last two are optional) must be filled to set the problem public and be accessible by other users.

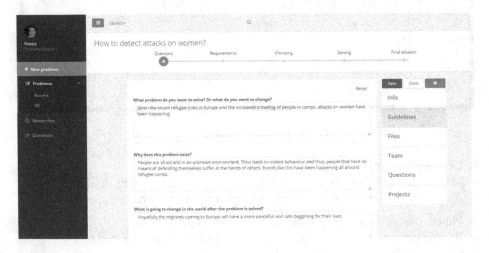

Fig. 4. Editing a problem

Files can be shared with team members and new members can be added to a problem. Another component that can be managed in the problem edit page, are the questions that

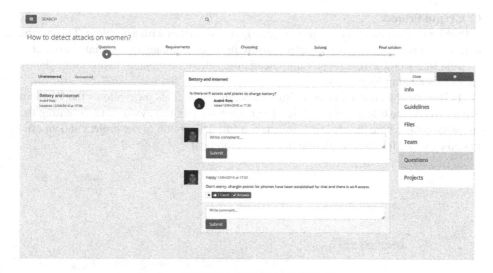

Fig. 5. Questions on problem edititng

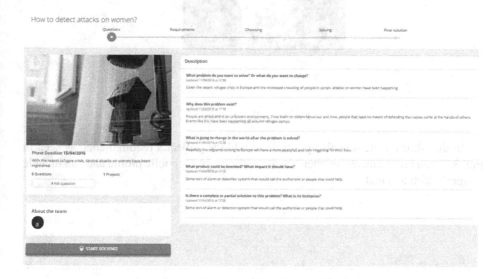

Fig. 6. Problem profile

are posted by users. In Fig. 5 questions are shown to be divided into answered and unanswered. A comment in an unanswered question can be marked as an answer and the question is moved to the corresponding list. The only users that can mark a comment as an answer are the Problem Makers in the problem's team.

When all the obligatory information is set, the problem can be set to public in the settings or editing pages.

The problem profile can now be seen by every user in the platform as it shown in Fig. 6. At the top of the page, the problem's current phase is indicated.

C. Create Project

A Problem Solver can only create a project if it is associated with a problem. This means that he first needs to access a problem profile in order to create an associated project.

Just like a problem, a project only needs the title to be created, being immediately set to private after its creation.

Following the creation is the editing page, in Fig. 7. This page allows the Problem Solver to edit the title and type a brief description (or pitch), keywords representing the subjects/fields where the project is inserted, a full description of the project and an estimated deadline for the completion of the development.

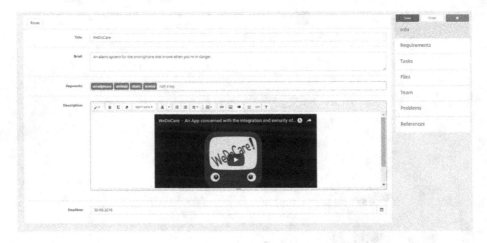

Fig. 7. Editing project

In the editing zone it is also possible to add the project's requirements, as seen in Fig. 8. These requirements are then shown in the project's public profile and can include images, video and text.

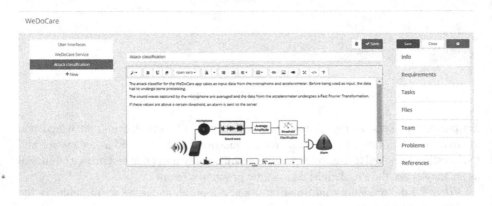

Fig. 8. Project requirements

The Problem Solvers working on the project can also manage the tasks involved in its development, like it is represented in Fig. 9. These tasks have three states: to-do, ongoing and completed. They can also be dragged from one of these lists to the other.

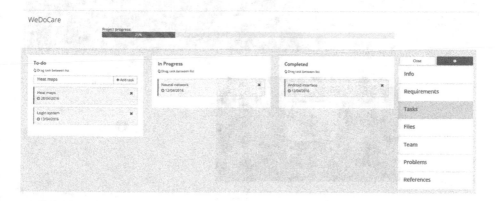

Fig. 9. Project tasks

A task needs a title. In fact, it is the only obligatory field; it also possible to add a description, assign one or more persons from the team to complete the task and define a due date.

Like a problem, files and new team members (Problem Solvers) can be added to the project. Also, the problems that the project is associated with (trying to solve) can be listed in the editing. Another component that can be listed are the researches being referenced by the project.

The project's state can be changed in the settings, although to make it public, the project needs a deadline (estimate) for its completion. Also in the settings it is possible to delete or leave the project.

If the project is public, then everyone can access its profile (Fig. 10). Here, the project's profile image (which can be set in the files area in the editing page), title, description, keywords, team and requirements are presented to the other users of the application.

A Problem Specialist not belonging to the project's team can recommend the project as a solution, through the project's profile.

D. Create a Research

Any user can create a research by only typing the title of the research. After a research is submitted, its editing area (shown in Fig. 11) is made available, allowing the change and input of new information. Title, description and keywords to use in the search for this article can also be changed in this page. The research has two states, public and private. By default, the latter is enabled.

Fig. 10. Project profile

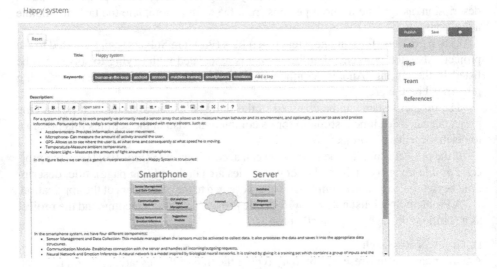

Fig. 11. Editing research

Files can also be uploaded to a research, although, contrary to problems and projects, these files are shared with the community and are considered attachments to the real research developed by the user.

Similarly, to problems and projects, new users can be added to the team of the research. Projects that referenced this research are also listed in the editing area.

When a project is set to public, its profile is provided and shown to other registered users. If the user accessing it is a Problem Solver, then an option to reference a particular

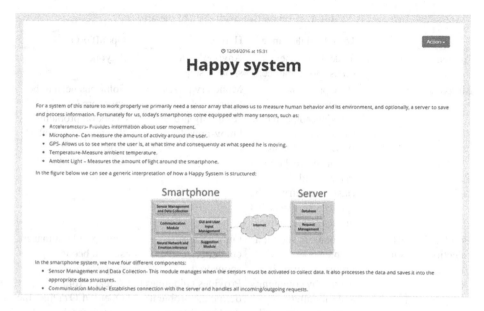

Fig. 12. Research profile

research is enabled. Comments by other users can be left in the research's profile (Fig. 12).

3.2 Comparison

Table 1 shows how similar aspects of two different platforms work in comparison to Tech4SocialChange. Even though others were studied, only two were included due to being the ones closest to the context of Tech4SocialChange.

HeroX allows for any sponsored entity to create a challenge/competition, which is normally based on a problem that the entity is currently facing. It provides management and counselling along the way for a certain fee. Also, to participate or access a challenge the user needs to pay, this ensures commitment and also helps in covering the costs of the competition.

OpenIDEO lets anyone submit a challenge or a social problem, to be evaluated and defined more clearly, ensuring that only problems that are sure to become projects are released to the community. Moreover, the selection process of solutions is based on OpenIDEO team's skills. This might result in a great solution being discarded.

Tech4SocialChange is aimed at a different audience, which is interested in the problem's context and the experience and recognition to be gained. Anyone can participate in submitting, defining and solving a problem. Even though prizes aren't awarded when a problem is solved, the Problem Solvers are given full recognition from Tech4Social-Change's side.

Table 1. Comparison with state of the art

	Tech4SocialChange	HeroX	OpenIDEO
Audience	Students, teachers, geeks and researchers	Anyone with sponsorship	Anyone
Motivation of use	Have problems tackled by experienced people in various areas Work on problems that have a real impact in the world Have research work be applied in projects that impact people's lives	Monetary prizes Sharing of competition-based know-how, for a fee	Solutions need to be sponsored Work on problems that affect the real world
Problem/Project selection and support	Anyone can submit a problem Problem criteria is defined by community and submitters Submitters choose the final solution; they can ask for help that is provided by Problem Specialists Projects only advance to development after they have been chosen. This prevents wasting time and resources building a solution that is not used after	Only accepts challenges that are sponsored Provides guidelines to define the problem An HeroX team manages the competition HeroX team helps in the choosing of a solution. along with the creator of the competition	Only problems that are sure to become projects A team from OpenIDEO defines the criteria for the problem The submitter and the team of OpenIDEO choose the solution according to the challenge criteria and the skills needed to build the solution
What happens when a solution is chosen?	Delivered to the NGO or other entity that submitted the problem Projects can be further developed and used on other problems	Prizes are awarded to the winner and project stays idle	Solutions are delivered to the OpenIDEO team that builds it and delivers it to the problem submitter. The solvers stay idle as well as the project

Some additional features are also provided based on the audience:

- All problems and projects are stored in the system and are accessible at all times. Projects can be re-used in different problems.
- Projects that have not been chosen do not stay idle; Problem Solvers can further develop them and present them on other problems.

- A research repository is also provided, letting researchers and scholars submit their work where it can be used and referred in projects and receive comments or ideas that may contribute towards its further development.

However, there are some concepts from problem solving that were adapted from other platforms into Tech4SocialChange.

For example, OpenIDEO's methodology of using skills people have that would otherwise get overlooked by the standard/popular problem-solving methods was taken advantage of. This concept allows for the final solution to be emotionally meaningful and also functional as it integrates feeling, intuition and inspiration with rational and analytical support. There are three concepts to keep in mind: inspiration (is the problem or opportunity that motivates the search for solutions), ideation (process of generating, developing and testing ideas) and implementation (path that leads the project to the real world). The three concepts motivated the creation of the different phases that a problem undergoes until a solution is found.

Also, HeroX's understanding of problem criteria was used to learn the key guidelines on how to better define/explain a problem, previously mentioned in Sect. 1.

4   Innovative Research Issues

As it is important to make a link between the researcher's work and real-world problems, the same link may also be applied a company's projects. As such, a search engine that makes a matching between research, companies and problems is necessary. A first version of this function is being developed using ElasticSearch - an open-source, scalable, full-text search and analytics engine. It allows to store, search and analyze big volumes of data and is highly used in applications that have complex search requirements.

At a structural level, the matching is done with simple text, matching the titles and keywords that represent the areas or subjects that a certain project/research/problem is inserted on.

Another important research venue is that of the intellectual property of the projects developed. Currently, the projects that are submitted to Tech4SocialChange are completely open-source and anyone can make use of the information and the products made available by the Problem Solvers. Those that do not wish for others to access their work can do so by setting a project as either "private" or "draft" (the first option shares it only with other teams that work on target problem, the second shares it exclusively with the team of the project itself). However, this might not be ideal for companies that wish to participate in Tech4SocialChange. As such, improved models of intellectual property will be object of future study and applied to the prototype.

A related research objective is understanding the language that both companies, entrepreneurs and NGO's have and create a bridge between them: what terms and visual aids can be used to minimize the gap between these three?

There is an interesting approach, presented in [8], concerning web-based collective design platforms. These platforms make use of their community to design and build solutions. OpenIDEO and another platform, Quirky, are studied to determine the main values that such a platform needs in order to motivate users and enhance the quality and diversity of solutions that are built. Tech4SocialChange is such a platform and since the study used, as case study, one of the platforms in the state of the art, it is interesting to determine how these values apply in Tech4SocialChange, which focuses on the academia and social problems, and perhaps further improve these values and/or set new ones.

Another objective is creating a model of specialization in different areas. What this means is that Problem Specialists are not only people specialized in problem definition but can also be specialized in different areas, or subjects, and be recognized as such. This way, projects can receive support from users that have knowledge about their specific subjects. It is a way of introducing help to inexperienced people (students for example) from others with greater experience on the field. This results in both better project results and greater learning experiences for students.

To determine and select the experienced people (Problem Specialists), a points system is currently being developed, and refined, based on the events, actions and achievements of the users (e.g. having a project chosen as a solution).

Nowadays people's interests and lifestyle are increasingly more integrated into the Internet: the Internet Of Things (IoT) uses low-cost technology that has a high potential of solving people's everyday issues in an non-intrusive way. Using this information to better match Solvers and Specialists to problems and researches that have a bigger connection with them is another research objective. Motivate people to work on subjects that interest them more. This would obviously be a major asset to problem solving and help tackling the challenges presented by Problem Makers.

5   Conclusions

There is already a prototype of the application that can be accessed in www.tech4social-change.org. It was developed by a team of students and researchers of the Department of Informatics Engineering of the University of Coimbra.

By allowing social problems to reach the academic world, we intend to not only solve them but also to approximate universities and institutions to real world situations and create many opportunities and cases of big impact in people's lives, in various parts of the world.

The next steps involve changes according to feedback being received by people that are helping in testing the prototype and also better support to tracking the contribution of people involved in problems, projects and researches.

An important link that it is hoped to be established is with entrepreneurship. How can Tech4SocialChange help the growth and establishment of entrepreneurs by finding situations in which their ideas and projects can be applied?

Acknowledgements. The work presented in this paper was partially financed by Fundação para a Ciência e a Tecnologia and POPH/FSE, as well by SENESCYT – Secretaría Nacional de Educación Superior, Ciencia Tecnología e Innovación de Ecuador.

References

1. Herox faq, January 2016. https://herox.com/faq
2. Herox about, January 2016. https://herox.com/about
3. Openideo faq, January 2016. https://challenges.openideo.com/faq?_ga=1.198164527.18040 87172.1444396793
4. Openideo about, January 2016. https://www.ideo.com/about/
5. Innocentive solver, January 2016. http://www.innocentive.com/faq/Solver
6. Codeforamerica fellowship, January 2016. http://www.codeforamerica.org/about/fellowship/
7. Hack4good, January 2016. http://hack4good.io/
8. Hajiamiri, M., Korkut, F.: Perceived values of web-based collective design platforms from the perspective of industrial designers in reference to Quirky and OpenIDEO. ITU AZ **12**(1), 147–159 (2015)

Topic and Object Tracking

Topic Tracking in News Streams Using Latent Factor Models

Jens Meiners[✉] and Andreas Lommatzsch[✉]

DAI-Labor, TU Berlin, Ernst-Reuter-Platz 7, 10587 Berlin, Germany
{jens.meiners,andreas.lommatzsch}@dai-labor.de

Abstract. The increasing number of published news articles and messages in social media make it hard for users to find the relevant information and to track interesting topics. Relevant news is hidden in a haystack of irrelevant data. Text-mining techniques have been developed to extract implicit, hidden information. These techniques analyze big datasets and compute "latent" features based on implicit correlations between documents and events. In this paper we develop a system that applies latent factor models on data streams. Our method allows us detecting the dominant topics and tracking the changes in the relevant topics. In addition, we explain how the extracted knowledge is used for computing recommendations based on trending topics and terms. We evaluate our system on a stream of news messages published on the micro-blogging service TWITTER. The evaluation shows that our system efficiently extracts topics and provides valuable insights into the continuously changing news stream helping users quickly identifying the most relevant information as well as current trends.

1 Motivation

With the fast growing volume of data available online, it becomes difficult for users to find and track the relevant information. The problem is very important in the news domain due to the huge number of published news articles and messages in social networks. Traditional newspapers printed once per day are replaced by news portals and blogs continuously releasing new articles. Social media also gains importance as it allows normal users to publish news articles and to give feedback on publications. Hence, the number of relevant sources as well as the frequency of published data increases.

The overload of data makes it hard for users to retain an overview of the relevant information, because relevant information is often hidden by irrelevant data. This intuition was confirmed by a survey conducted on young news consumers by Associated Press in 2007 [1]. The study shows that although more information is available, users do not gain deeper insights into political subjects: *"Participants in this study showed signs of news fatigue; that is, they appeared debilitated by information overload and unsatisfying news experiences"* [2]. Carpini et al. stated that, *"As choice goes up, people who are motivated to be politically informed take advantage of these choices, but people who are not move away from politics"* [2].

© Springer International Publishing AG 2016
G. Fahrnberger et al. (Eds.): I4CS 2016, CCIS 648, pp. 173–191, 2016.
DOI: 10.1007/978-3-319-49466-1_12

The overall conclusion is that readers need more context and coherence to handle the oversupply of information.

1.1 The Analyzed Problem

In order to overcome the data overload problem, tools are needed supporting users in finding the relevant information in streams. Important subtasks are the identification of relevant topics and the tracking of these topics.

Most of the existing solutions for the analysis of news data are built using static datasets. These approaches do neither support the tracking of topics nor the detection of trends. The efficient processing of streams is a challenge because with continuous change and the large amount of new data in streams, an efficient scaling with the volume and the velocity of the analyzed streams is required.

1.2 Our Contribution

In this paper we present an approach optimized for the continuous extraction of the dominant topics from a stream of documents. We develop an algorithm for tracking the topics and computing trending topics and terms. The identified trends help users finding relevant content and understanding the evolution of topics. The tracking of topics and the visualization of correlations between them leads to a better understanding of news. Hence, the overall information presented to the user is reduced to the most descriptive minimum. In order to ensure a real-time processing of large news streams, our approach addresses the scalability of the text-mining task by applying the Map-Reduce paradigm enabling the efficient execution in distributed environments.

1.3 Structure of the Paper

The remaining paper is structured as follows. Section 2 describes the scenario and explains the requirements. In Sect. 3 we discuss state-of-the-art approaches related to the analyzed problem. Based on existing techniques, we develop our approach in Sect. 4. We present the system's architecture and explain the components in detail. Section 5 describes the evaluation of our system. We discuss the strengths and weaknesses of the algorithms with respect to the analyzed scenario. Finally a conclusion and an outlook to future work are given in Sect. 6.

2 Problem Description and Dataset

Driven by the ubiquity of internet connectivity and the growing importance of social media and blogs, the amount of news published each minute increases. In addition, users always want to be up-to-date; new topics and new trends are highly relevant for most users. Due to the huge amount of data it is difficult for users to extract the relevant information from the large volume of continuously released news.

2.1   The News Recommendation Scenario

In this work we focus on the analysis of news articles published on TWITTER. TWITTER is a popular micro-blogging service allowing users to post text messages (so called "tweets") that have a maximum length of 140 characters. Evan Williams (CEO of TWITTER) defines the service as follows: *"What we have to do is delivering to people the best and freshest most relevant information possible. We think of Twitter as it's not a social network, but it's an information network. It tells people what they care about as it is happening in the world"* [3].

The messages published on TWITTER are very diverse, but there is a strong bias towards news. Kwak et al. [4] state that 85 % of all tweets are based on current news headlines, hence TWITTER is a very prominent source for collecting the most recent news data.

For our analysis we created a dataset consisting of news tweets from the most important German newspapers, such as FAZ, SZ and ZEIT. The messages have been collected using the twitter streaming API[1]. This approach ensures that we get a large stream with the most recent news published by trustful authors. A special filter for spam or meaningless status messages is not needed. The diversity and the reasonable number of considered newspapers ensure that we cover all relevant news domains and get enough messages for a deep analysis.

News Stream. In contrast to static datasets, data streams inherently have a dynamic nature. Characteristic properties of streams are its volume, velocity and veracity. The stream dynamics can be observed with respect to content aspects as well technical aspects.

Regarding the content of streams, changes in the set of entities and the related properties are especially important. In the analyzed news scenario, this means the occurrence of fresh articles and the removal of outdated articles as well as changes in the set of occurring terms and the evolution of topics over time. Furthermore, the number of retweets may change over time. On the technical side, alternating volume and velocity of streams are the key aspects that the stream-based recommender system must take into account. In our news scenario, unexpected highly engaging events often result in load peaks created by news updates on the topic and huge number of user interactions (e.g. retweets).

We address these issues by applying two approaches: (1) Heuristics are used for estimating the total number of retweets based on the retweet count for a tweet in the first 60 min. The time range was chosen based on statistics on the dataset and the lifespan of tweets [5]. Thus, our tweet collection strategy does not only collect new tweets, but also updates the meta-data for already known tweets. (2) The challenge that arises from the significant change in data volume per time frame is addressed by a map reduce paradigm that can be distributed across a reasonable number of nodes. Our component enables us collecting up to half a million characters per minute (not including additional meta-data). All tweets are processed as a stream that is generated continuously.

---

[1] Twitter streaming API: https://dev.twitter.com/streaming/overview.

Table 1. The table shows the top 5 publishers (based on the total tweet count) and the key figures describing the volume of published data between June 1st, 2015 to March 1st, 2016. The publisher's names are the so called TWITTER *short names*.

Publisher (short user-name)	Tweet count	Max retweets	Average retweets	Sum retweets
WELT	33,435	305	9.45	315,941
FAZ_NET	27,796	68	1.06	29,538
SZ	17,034	224	7.61	129,576
SPIEGELONLINE	14,629	462	13.49	197,269
TAGESSPIEGEL	14,605	2,991	6.48	94,702
Others	132,751	1,150	6.87	911,422

Dataset. The dataset of collected tweets is based on the tweet stream of several popular German news publishers. It contains approximately 240,000 documents (from the June 1st, 2015 to the March 1st, 2016). Table 1 lists the top 5 publishers with most tweets in the analyzed time frame. Note the large differences in the average retweet counts and outliers in the maximum retweet counts. Table 2 lists the top 10 hashtags and the number of retweets of the tweets containing these hashtags. The table shows that the average retweet count differs a lot from the total tweet count. Based on the number of retweets, the TWITTER messages containing the hashtags #flüchtlinge and #türkei are much more relevant than tweets relating to #berlin or #polizei. This shows that considering retweets gives an alternative view on reported news.

Table 2. Top 10 hashtags regarding total tweet count with respective maximum and average retweet count from the June 1st, 2015 to March 1st, 2016.

Hashtags	Tweet count	Max retweet count	Average retweet count
#BERLIN	13,151	348	1.85
#FLÜCHTLINGE	4,160	1,396	13.41
#GRIECHENLAND	3,384	415	6.55
#POLIZEI	2,997	155	1.69
#STUTTGART	2,110	35	2.06
#SYRIEN	1,555	162	8.84
#MERKEL	1,274	243	9.71
#IS	1,197	162	9.55
#EU	1,156	165	7.22
#TÜRKEI	1,128	165	13.38
Others	212,056	2,991	7.18

2.2 Requirements and Challenges

In order to support users in finding the most relevant topics in the news stream, a set of content-based requirements and technical demands must be fulfilled. In this section we discuss the requirements and explain the related challenges.

Identifying Topics. The system should extract the most important topics from the document stream. The analysis should consider implicit synonyms and co-occurring terms. A latent factor analysis is a promising approach for a deep analysis enabling the identification of the most important topics and extraction of hidden characteristics for the detected topics.

Analysis of Continuous News Streams. The system must be able to process continuous streams. In contrast to existing systems handling static sets, our focus is on the processing of streams characterized by high velocity and steadily changing volume. The algorithms used for the latent factor analysis as well as the tracking of the most relevant topics must be adapted accordingly.

Topic Tracking and Trending Detection. The system should be able to track identified topics in streams over time. This means that the system should compute trending topics as well as trending terms helping the user to find new relevant aspects.

Compute Predictions and Recommendations. Recommendations help users in finding relevant items. News streams provide a huge amount of data that tends to surpass the users consumption capacities. Our system should be able to extract relevant information from the stream and compute recommendations supporting users in discovering interesting terms and subjects.

Openness and Support of Heterogeneous Sources. The system should focus on the efficient processing of news streams. In order to ensure the openness and extensibility of the system, the approach should use a flexible, abstract data representation. News documents are processed using large matrices describing the relation between two entity types. This model can be adapted by integrating additional relationships that can be represented in matrices. This ensures that the developed approach can also be applied in other domains, e.g. for tracking trends in social networks.

Scalability. The fast processing of complex data requires an approach that scales well with the amount of input data. The approach should be executable in distributed environments (e.g. cloud-based computing environments). This allows the operator adding nodes and resources with respect to the data volume and the time constraints.

Systems developed for extracting and tracking topics in news streams must fulfill complex requirements. Before we present our approach, we review related algorithms and systems in the next Section.

3   Related Work

The clustering of documents and the detection of "topics" in static document sets are well-known tasks in Information Retrieval [6]. Several different approaches for these tasks exist, such as *dimensionality reduction* [7], probabilistic clustering [8], and graph-based methods [9]. In this Section, we review current topic detection algorithms. We focus on algorithms enabling the handling of massive amounts of data and the tracking of topics in streams. In addition, we analyze existing systems for topic tracking in news streams and discuss the addressed use cases as well as the applied algorithms.

3.1   Topic Detection Models

Topic detection approaches typically incorporate latent feature models [10,11]. Documents are tokenized and treated as a bag of terms. The basis for the latent feature analysis is a document-term matrix handling the term distribution within as the features describing the characteristics of the documents. The use of matrix representations allows the efficient processing of large document collections taking into account the sparsity of the term-document matrices. However, the processing of large document collections is computationally expensive. This means that the processing is still slow and cannot be done on the fly. In order to overcome this issue, several incremental analysis algorithms have been developed applying latent feature models on streamed data in an online fashion [12–14]. In the following paragraphs, we discuss different approaches for detecting latent factor models with a focus on stream-based scenarios.

Non-negative Matrix Factorization: Cao et al. [12] modified an dimensionality reduction model (NMF) for an online scenario such that single documents can be incorporated into the final results without revising old data. They evaluated the algorithm on nearly 8,000 news articles in a timeframe of one month. A supervised set of terms has been applied for describing the observed topics. Therefore, the online capabilities of topic detection are not depicted in the evaluation. Another problem is that the dimensionality reduction was transposed into a minimization problem that may get stuck in local minima. Hence, these approaches do not necessarily find the optimal solution. Vaca et al. [13] also use a NMF approach but focus explicitly on topics and how they evolve over time.

Latent Dirichlet Allocation: The most popular approach to latent factor analysis is LDA [15]. Alsumait et al. [16] modified the model to be applicable in online scenarios. The advantage of this approach is that correlations and dependencies of topics are handled. A weakness of LDA is that a probability distribution is assumed might not suitable for the fast changes when analyzing news. Wang et al. [17] leverage the model to learn the relationships or transitions among topics. While previous algorithms learn word distributions for a topic, their approach tries to model probabilities of topic transitions. The suggested models work in an online fashion making it hard to determine the number of relevant topics as well as optimizing the required hyper parameters [13]. Batch-based models are designed to overcome this problem.

3.2 News Recommender Systems

Recommender systems help users finding interesting items in huge collections of data. With the growing popularity of online news in the last decades several recommender systems for the news domain have been developed. Balanović et al. [18] developed a system combining collaborative and content-based algorithms for recommending interesting web pages. Liu et al. [19], Morales et al. [5], and Li et al. [20] present recommender systems optimized for the news domain. These systems compute the recommendations based on user profiles that contain keywords describing the user preferences. The disadvantage of this approach is that recommendations are limited to topics the user is already familiar with. The user is caught in a "filter bubble"; new trends and topics are often not provided in the recommendation results. In addition, it is often difficult for users to find the optimal keywords precisely describing the individual user preferences.

3.3 Identifying and Tracking Topics

Recommending news is a hard task due to the heterogeneity of news sources and the continuous changes in the world. An approach for aggregating news from different sources is the detection of topics using latent factor models. Saha et al. [21] use a NMF model for extracting topics from a news stream and analyzing term-document matrices. Hennig et al. [22] present a system that identifies topics by clustering Named Entities in multi-lingual news. The tracking of the evolution of topics is implemented based on the Vector Space-based similarity measure. New news article having a similarity to an existing topic below a threshold are added to an existing cluster; otherwise a new cluster is created. The disadvantage of the incremental creation and adaptation is that these approaches only slowly adapt to new topic distributions. This makes it difficult to identify trending topics and computing recommendations.

3.4 Discussion

Latent Factor models implemented using LDA and NMF have been successfully applied for clustering document collections and for mining dominant topics. Most of the systems focus on static datasets (e.g. [7]) or use a controlled vocabulary for mapping documents on popular clusters (e.g. [12]). These systems show that latent factor models are an appropriate method for static collections.

News recommender systems focus on computing the similarity of news articles with user profiles. The disadvantages of this approach are that articles belonging to new trending topics are not considered as relevant. In addition, users must maintain the user profile ensuring that only relevant terms and their synonyms are kept in the profile.

Systems designed for topic detection are mostly built on clustering algorithms developed for static sets. The handling of streams is not supported. Systems for tracking news are usually based on explicit user profiles or implicit profiles computed based on cluster algorithms. The extension of latent topic models to be capable of handling streams seems to be a promising approach. The additional complexity arising from the processing of streams instead of static sets is manageable when applying distributed algorithms and modern machine learning frameworks.

4   Approach

In this section we present the architecture of our system and describe the developed components in detail. We explain the implemented algorithms and discuss how the approach addresses the identified challenges.

4.1   The System Architecture

One of the key requirements of the designed system is the efficient processing and analysis of heavy data streams. In order to enable the distributed crawling of data and the scalable data analysis we implement a 3-layer architecture. The crawling and the data analysis are decoupled.

The data is cached using an ELASTICSEARCH[2] index created by the crawler components. The crawled data is the input for the analysis layer. The analysis results are stored in data cubes persisted in a database. For the visualization of the results we implement a web GUI and a component periodically generating newsletters aggregating the relevant information (Fig. 1).

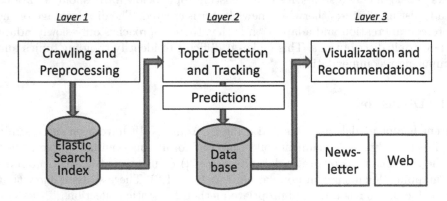

Fig. 1. The figure visualizes the 3-layer architecture. The processing nodes are decoupled by data stores implemented using ELASTICSEARCH and databases.

---

[2] https://www.elastic.co/products/elasticsearch/.

In order to ensure the scalability of the processing and analysis components, our architecture supports the use of powerful distributed frameworks such as APACHE SPARK or APACHE FLINK. These frameworks simplify the use of advanced machine learning algorithms in distributed environments and allow us adding new computational resources if needed.

4.2   Crawling and Preprocessing

The crawlers fetch the data provided by the TWITTER API. The API gives access to all new tweets starting after a given `tweetID` or all tweets matching a set of identifiers. While the first approach is used for continuously tracking the stream of news, the second method is used for updating the retweet count for already collected tweets. New tweets are preprocessed before persisting them in the index. We use the ELASTICSEARCH framework supporting the distribution of the index on distributed nodes. The preprocessing transforms the raw data into a representation optimized for the later analysis. We apply the natural language processing methods, such as encoding detection, language detection, stop word removal, and stemming.

4.3   Topic Tracking and Predictions

We discuss approaches for detecting topics, tracking trends, and computing predictions.

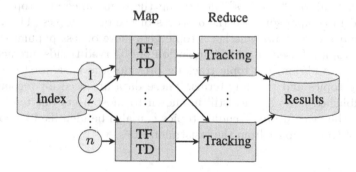

Fig. 2. The figure visualizes the map-reduce paradigm applied to the generation and analysis of timeframes (TF). Latent factor models are used for detecting the topics (TD). Data is assembled from different, separated data nodes $(1, 2, \ldots, n)$.

As basis method for analyzing the news stream we apply a latent factor model for topic detection. We segment the news stream into timeframes and represent each timeframe as a term-document matrix containing the tf-idf scores of the documents. As a result of the latent factor analysis, we obtain term vectors describing the identified topics based on a set of weighted terms and a score describing the dominance of the topic in the timeframe. The size of the frames has a strong influence on the results and must be chosen dependent on the specific source. In the analyzed scenario we focus on timeframes having a size of 6 to 24 h. This seems reasonable in order to catch trends of the relevant phases of the day and ensuring that there are enough documents in each timeframe for a reliable analysis.

The document-term matrices created for a timeframe are analyzed by applying a Singular-Value-Decomposition (SVD). Since we make use of the Map-Reduce paradigm, the analysis is done in the Map-phase. In order to track the changes in identified topics, we use a sliding window approach. We shift the window so that the new window overlaps with the already analyzed windows by 80 % ensuring that we can re-detect the "old" topics enabling us studying the changes in the topics. For matching the "old" topics with the "new" topics, we use the cosine similarity. Since the term-vectors describing the latent topics are normalized, this metric is defined by the dot product of the term vectors enabling the efficient calculation of the metric. The result of this phase is a mapping of topics based on the topic vectors of the different timeframes. This allows us detecting and tracking topics as well as analyzing the shifts inside the topics. The data of the analysis are stored in a database giving us flexible access for a deeper analysis and the possibility of later visualization of the information.

4.4 Visualization and Recommendation

Based on the extracted latent topics and the related term vectors different visualization approaches can be used for guiding users in finding useful information. The singular values ("σ scores") describing the dominance of a topic must be interpreted in comparison with the σ scores of the other topics. The visualization of the σ scores shows whether a topic gets more or less popular compared with other topics. Based on an extrapolation of observed trends, predictions are computed. In addition, the topic tracking provides the basis for the detection of trending topics and trending terms. These data are used for recommending freshly published news articles still unknown to most users but tending to be relevant in the near future. Trending topics can also be identified by comparing the slope of topic impact from the last timeframes.

4.5 Scalability

The architecture is designed for supporting data sources providing a huge volume of data. For ensuring scalability, data streams can be split in time windows that can be processed concurrently. The (ELASTICSEARCH) index used for decoupling the information sources and the processing supports the distribution on several machines.

Caching the documents in the index enables the concurrent distributed processing of data. To scale with a fast growing index, the map reduce paradigm is applied (Fig. 2). The map tasks aggregate the timeframes from the index nodes (might running distributed on different machines). Within the cluster, all nodes are queried in order to assemble a specific timeframe, but only the ones relevant retrieve the data from their hard drives. In the reduce phase, coherent timeframes are gathered for the topic tracking to be applied to. The results of this task are persisted in a database for further analysis and visualization tasks.

5 Evaluation

In this Section, we discuss the evaluation of the developed system. We analyze the precision as well as the scalability of the topic detection and tracking models.

5.1 Latent Factor Models

We evaluate the detection of topics from June 1st till August 31st, 2015. Figure 3 visualizes the three most dominant topics per day. The graph shows that topics stay in the top 3 topics list for one to four days. Typically there is one topic per week listed for 3–4 days. Most of the topics have a short life cycle; these topics stay only one day in the top 3 list.

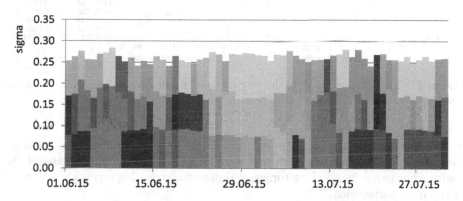

Fig. 3. The diagram shows the changes in the top three topic distribution over time if re-tweets are not considered.

Figure 4 visualizes the topics on the same timeframe. In contrast to the previously discussed dataset, the documents are weighted based on the number of retweets. The document weight is assigned linear with the number of retweets. The figure shows that the detected topics have a much shorter lifecycle. This underlines that users focus on retweeting new topics. Long-livings topics still fill the newspapers, but the user engagement, measured by number of retweets, is only high for new topics.

The differences in the dominance of latent topics are also visible when analyzing the weights of the top 15 topics. Figure 5 shows the σ scores for 14 days. The graph shows that considering retweets results in significantly higher weights for the most dominant topics. The σ scores follow a power law distribution.

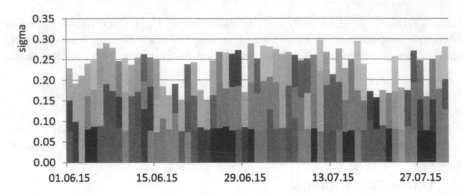

Fig. 4. The diagram shows the changes in the top three topic distribution over time if retweets are taken into account.

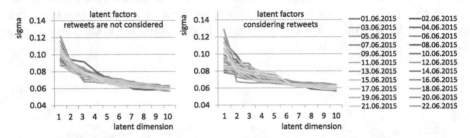

Fig. 5. The graphs show that considering retweets result in a greater variance in the σ scores describing the dominance of topics.

We showed that the continuous latent factor analysis provides the basis for the detection of topics based on a stream of news articles. The details of the topics, approaches for tracking topics, and detecting trending terms are discussed in the next paragraphs.

5.2   Tracking Topics

We study how the topics (identified based on latent factor analysis) develop over time. For this purpose we investigate for one selected topic how the term vector describing the topics changes.

How Fast Does a Topic Change. In order to analyze the evolution of a topic, we measure the changes of the term vector describing the centroid of each topic. The degree of change is measured by the cosine similarity between the term vector at timeframe i and the term vector at timeframe $i + 1$: $\text{sim}(v_i, v_{i+1}) = \frac{v_i \cdot v_{i+1}}{\|v_i\| \cdot \|v_{i+1}\|}$. A similarity score of 0 indicates that both vectors do not have common terms and the topic centroid changes strongly. A similarity score close to 1 indicates that both vectors are similar and the topic does not change.

Figure 6 visualizes the similarity scores computed for one selected topic for the first 2 weeks in June 2015. The graphs show, that there are no regular patterns. At several days the similarity score is below 0.2 indicating strong changes in the topic.

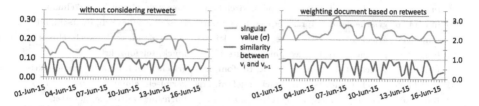

Fig. 6. The graphs visualize the singular value and the changes of a selected topic for the first two weeks of June 2015.

Comparing the two graphs, the first computed without considering retweets and the second considering retweets, we find that the graph changes much stronger if retweets are taken into account. A trend enforced by retweets lasts only one day on average. At the following day, the topics adapt to a new news event. Analyzing the topics without retweets it can be observed that the changes are more smoothly.

In order to get deeper insights in the evolution of a topic, we analyze the dominant terms and the changes in the term vector over one week. Table 3 shows the terms having the highest impact on the topic. In addition, we compute the most important terms added to the topic. The presented evolution of the topic shows, that trending terms tend to be the most dominant terms for a topic. This means that previously unobserved terms receive a lot of attention in the first days. There is no obvious pattern describing how long a term is highly relevant for a topic.

5.3   Visualization of the Evolution of Topics

Our recommender extracts topics from the stream of news articles. A slightly adapted version of the implemented approach can also be used for visualizing the evolution of topics. We compare two approaches. (1) We compute how dominant selected topics are present in the news stream. The dominance is measured by projecting the news stream in a one-dimensional space spanned by the topic term vector. (2) We define a two dimensional space spanned by the two selected topics. We project the stream of documents into this space. Figure 7 visualizes both methods for the topics refugees and greece. The left part of the figure shows the dominance depending on the date. Both topics are highly relevant over the complete analyzed period, but the correlation is not visible in the graph.

The right part of Fig. 7 shows the news stream projected into a space defined by the two dominant latent topics. We extracted the topics in the latent factor analysis on June 1st, 2015. These topic term vectors define the axis of the space

Table 3. The table shows the evolution of a topic in the year 2015. For each time frame the most dominant terms and the trending terms are shown (sorted descending by their weight). The tables shows the words reduced to their stem.

Date	Dominant terms	Trending terms
06 Jun, 02:00	griechenland, tsipras, merkel, stuttgart	tsipras, turkei, schaubl
06 Jun, 08:00	griechenland, merkel, tsipras, kirchentag	garmisch, elmau, stopg7elmau
06 Jun, 14:00	griechenland, tsipras, leipzig, asylbewerb	leipzig, asylbewerb, deutschland
06 Jun, 20:00	griechenland, tsipras, asylbewerb, leipzig	deutschland, asylbewerb, elmau
07 Jun, 02:00	griechenland, tsipras, junck, deutschland	junck, telefoni, kris
07 Jun, 08:00	griechenland, tsipras, junck, telefoni, kris	griechenland, tsipras, schuldenkris
07 Jun, 14:00	griechenland, tsipras, junck, telefoni, kris	garmisch, elmau, obama
07 Jun, 20:00	fluchtling, mittelme, gerettet, kust	fluchtling, mittelme, gerettet
08 Jun, 02:00	turkei, turkei, akp, fluchtling	erdogan, akp, mehrheit
08 Jun, 08:00	turkei, erdogan, akp, wahl	akp, mehrheit, g7
08 Jun, 14:00	g7, turkei, gipfel, erdogan	g7, gipfel, merkel
08 Jun, 20:00	g7, gipfel, elmau, g7gipfel	g7, gipfel, klimaschutz

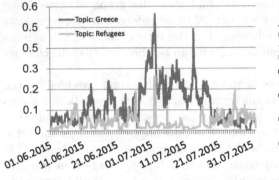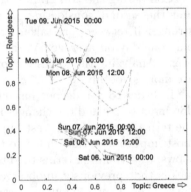

Fig. 7. The graphs visualize the evolution of two selected topics.

used for visualizing the evolution of the topics. For the visualization of trends we aggregate all documents published in a specific time window (having a size of one hour) and compute how this collection correlates with the topics (used as the basis of our sub-space). The correlation is measured using the cosine similarity with the selected base topics. Figure 7 visualizes the evolution of the topic. At the start point (Sat, June 6th) the news stream contains a lot of documents relevant to the topic Greece; the topic Refugees is also present but has only a small impact in the news stream. The graph shows, that at Monday, June 8th topic Refugees becomes dominant.

Both approaches visualize the dominance of the two topics over time. In the example we used a sliding-window step size of six hours. The use of a smaller step size smoothens the graph.

5.4   Recommendations

The analysis of the topics evolving over time helps us understanding the topics discussed on TWITTER. We use the detected patterns for generating recommendations by computing the terms freshly occurring in the term statistics. The applied latent factor analysis allows us extracting hidden information and detecting current trends. Recommendations created based on latent factors as well as on new, trending terms support users in finding fresh, potentially still unknown information.

Our analysis shows that the prediction what topics will be most dominant the next day works well. The detailed investigation of the lifecycle of topics (extracted when considering retweets) indicates that these topics have a very short time to live. The analysis of the news stream (without considering retweets) shows that the dominant topics are discussed over several days. In order to support users in tracking topics, we compute the terms newly added to the relevant topic. Recommendations based on these terms help users understanding current changes in the topics and identifying the new aspects relevant for a topic.

The investigation of trends is a powerful strategy for computing recommendations. Trends allow us suggesting new information (still unknown to the user). Since the recommendations are computed based on topics familiar to the user, there is a high chance that the recommendations are also relevant. In order to recommend previously unknown topics, the recommendations should be computed based on the trending topics (and the related terms).

5.5   Scalability

The presented system is designed to processes data streams in real time. For proving the scalability of our system we analyze the required processing time depending on the volume of analyzed data. All experiments have been run on a single machine having an Intel® Xeon® CPU (X7550) with 2.00 GHz and 4 GB of RAM.

Figure 8 shows that the processing time grows linearly with the number of analyzed days. The window size (visualized by the different blue lines) has only a small influence. Bigger windows result in a higher offset that can be explained by the fact that more tokens must be analyzed before starting with the processing of the first window. The deviation of the processing time from the expected linear correlation is induced by the variations in the number of published tweets. Analyzing the influence of the number of unique tokens on the processing time (right part of Fig. 8), we find that the processing time grows linearly with the number of unique tokens. On the machine used for the analysis the processing took ≈ 0.3 ms per token. This allows us to process the data of one day in ≈ 0.6 s.

For a deeper analysis of the scalability we used TWITTER FIREHOSE as data source. This source provides a stream of $\approx 350,000$ tweets per minute. Analyzing the run time of our system while processing that stream, we find that our approximation explained the previous paragraph holds true not only for tweets from German news-portals, but also for streams having a much larger volume.

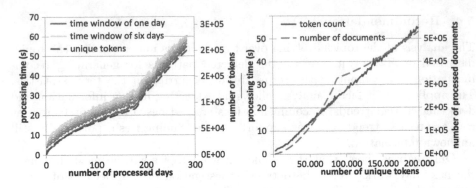

Fig. 8. The figure visualizes the processing time dependent on the volume of the data stream. Different shades of blue in the left subfigure indicate time window sizes of one to six days. For both plots, blue relates to the processing time while red relates to the second vertical axis. (Color figure online)

It takes about 5.8 min to process a single time frame on a single core machine. Therefore, six mapper tasks should suffice to keep up with the stream of data. In a production environment, the number of available mappers should be reasonably higher to keep up with radical changes in the volume of the stream. A higher number of mappers however cannot change the fact that it takes more time to finish a single window if the stream increases in volume. More mappers, each on its own computational unit, guarantee that computational power is still available for new time windows. As more data is streamed into the system, several mappers will take more time to process available windows and hence, will not finish as soon as new windows can be fetched. Therefore, the system needs to provide a reasonable amount of mappers that can take over new jobs. To conclude, the map reduce paradigm does not help in delivering results faster, but makes it possible for the system to keep up with the increasing amount of data while not falling back with more time windows becoming available. Also, since an increasing stream volume triggers an increasing use of computational power this system does indeed scale with the stream while the delay of available results may increase. To provide results even faster, there are three possibilities. (1) A more powerful CPU could help to overcome the limiting hardware. (2) We could optimize the latent factor model. Since the limiting factor is the matrix size describing the document term relation of each time window, another model may yield faster results in exchange for a reduced accuracy. (3) A third approach is the use of smaller time frames. Smaller time frames allow the mappers working on smaller data and finishing tasks faster. The main problem about this approach is that the topic detection may yield in a reduced precision if there are not enough diverse documents in a time frame to reconstruct the discussed topics. For this approach, a well-balanced topic distribution among the tweets is needed to guarantee stable results.

5.6   Discussion

The evaluation shows that the developed system fulfils the identified requirements. The continuous analysis of the news stream gives us interesting insights. In the news dominant topics change frequently. This makes it hard to predict trends and to compute recommendations. The analysis shows that the identification of trending topics and terms works and supports users in finding relevant information. We presented different approaches for visualizing the evolution of topics focusing on specific aspects of the news stream.

6   Conclusion and Future Work

In this paper we presented a novel approach for continuously detecting and tracking latent factors in news streams. In contrast to approaches based on term vector clustering, we compute the dominant topics based on the latent factors for each time window in the dataset. Our approach allows us efficiently identifying the topics and the most relevant terms.

The temporal analysis of clusters over time provides us with the trending terms from each topic and the terms relevant over a longer period of time. We showed that considering retweets typically results in short peaks indicating a high interest in topics. Since the peaks are short, retweets do not support the tracking of topics. The information about the lifecycle of the terms allows us computing recommendations based on trending topics and terms.

We showed that our approach scales linearly with the number of unique terms considered in an analyzed time window. The use of the map reduce paradigm enables the distributed processing on different cores or machines and therefore the scalability of our system.

Future Work. The current system extracts latent factors and trends from the news stream. The recommendations do not take into account individual preferences. As future work we plan to personalize the recommendations by assigning document weights based on user profiles.

Furthermore, we plan to apply our approach to additional domains. Our approach is based on matrices describing the relation between two entity types. These entity types can be easily adapted. E.g., the relationship between users and items in online shops or ratings for media could be modeled in the matrices. In this scenario the evolution of latent factors and the extraction of trends are relevant topics when developing services to support the user or when computing recommendations.

Acknowledgement. The research leading to these results was performed in the CrowdRec project, which has received funding from the European Union Seventh Framework Programme FP7/2007-2013 under grant agreement No. 610594.

References

1. Associated Press: A New Model for News: Studying the Deep Structure of Young-Adult News Consumption, July 2008
2. Nordenson, B.: Overload! Columbia J. Rev. **30** (2008)
3. Sprenger, T.O., et al.: Essays on the information content of microblogs and their use as an indicator of real-world events. Dissertation, Technische Universität München, München (2011)
4. Kwak, H., Lee, C., Park, H., Moon, S.: What is Twitter, a social network or a news media? In: Proceedings of the 19th International Conference on WWW 2010, pp. 591–600. ACM, New York (2010)
5. De Francisci Morales, G., Gionis, A., Lucchese, C.: From chatter to headlines: harnessing the real-time web for personalized news recommendation. In: Proceedings of the 5th ACM International Conference on Web Search and Data Mining, pp. 153–162. ACM (2012)
6. Manning, C.D., Raghavan, P., Schütze, H., et al.: Introduction to Information Retrieval, vol. 1. Cambridge University Press, Cambridge (2008)
7. Perkio, J., Buntine, W., Perttu, S.: Exploring independent trends in a topic-based search engine. In: Proceedings of the IEEE/WIC/ACM International Conference on Web Intelligence, pp. 664–668. IEEE Computer Society (2004)
8. Srivastava, A.N., Sahami, M.: Text Mining: Classification, Clustering, and Applications, pp. 121–180. CRC Press (2009)
9. Mei, Q., Zhai, C.: Discovering evolutionary theme patterns from text: an exploration of temporal text mining. In: Proceedings of the 11th International Conference on Knowledge Discovery in Data Mining, pp. 198–207. ACM (2005)
10. Xu, W., Liu, X., Gong, Y.: Document clustering based on non-negative matrix factorization. In: Proceedings of the 26th International ACM SIGIR Conference on Research and Development in IR, pp. 267–273. ACM (2003)
11. Deerwester, S.C., Dumais, S.T., Landauer, T.K., Furnas, G.W., Harshman, R.A.: Indexing by latent semantic analysis. JAsIs **41**(6), 391–407 (1990)
12. Cao, B., Shen, D., Sun, J.-T., Wang, X., Yang, Q., Chen, Z.: Detect and track latent factors with online nonnegative matrix factorization. In: IJCAI, vol. 7, pp. 2689–2694 (2007)
13. Vaca, C.K., Mantrach, A., Jaimes, A., Saerens, M.: A time-based collective factorization for topic discovery, monitoring in news. In Proceedings of the 23rd International Conference on World Wide Web, pp. 527–538. ACM, New York (2014)
14. Brand, M.: Incremental singular value decomposition of uncertain data with missing values. In: Heyden, A., Sparr, G., Nielsen, M., Johansen, P. (eds.) ECCV 2002. LNCS, vol. 2350, pp. 707–720. Springer, Heidelberg (2002). doi:10.1007/3-540-47969-4_47
15. Blei, D.M., Ng, A.Y., Jordan, M.I.: Latent Dirichlet Allocation. J. Mach. Learn. Res. **3**, 993–1022 (2003)
16. AlSumait, L., Barbará, D., Domeniconi, C.: On-line LDA: adaptive topic models for mining text streams with applications to topic detection and tracking. In: Proceedings of the 8th IEEE International Conference on Data Mining, ICDM 2008, pp. 3–12. IEEE (2008)
17. Wang, Y., Agichtein, E., Benzi, M.: TM-LDA: efficient online modeling of latent topic transitions in social media. In: Proceedings of the 18th ACM SIGKDD International Conference on Knowledge Discovery and Data Mining, pp. 123–131. ACM (2012)

18. Balabanović, M., Shoham, Y.: Fab: content-based, collaborative recommendation. Commun. ACM **40**(3), 66–72 (1997)
19. Liu, J., Dolan, P., Pedersen, E.R.: Personalized news recommendation based on click behavior. In: Proceedings of the 15th International Conference on Intelligent User Interfaces, pp. 31–40. ACM, New York (2010)
20. Li, L., Li, T.: News recommendation via hypergraph learning: encapsulation of user behavior and news content. In: Proceedings of the 6th ACM International Conference on Web Search and Data Mining, pp. 305–314 (2013)
21. Saha, A., Sindhwani, V.: Learning evolving, emerging topics in social media: a dynamic NMF approach with temporal regularization. In: Proceedings of the 5th ACM International Conference on Web Search and Data Mining, pp. 693–702. ACM (2012)
22. Hennig, L., Ploch, D., Prawdzik, D., Armbruster, B., Düwiger, H., De Luca, E.W., Albayrak, S.: SPIGA - multilingual news aggregator. In: Proceedings of GSCL 2011 (2011)

Collaboration Support for Transport in the Retail Supply Chain

A User-Centered Design Study

Marit K. Natvig[✉] and Leendert W.M. Wienhofen

Software Engineering, Safety and Security, SINTEF ICT, Trondheim, Norway
{Marit.K.Natvig,Leendert.Wienhofen}@sintef.no

Abstract. The purpose of this paper is twofold. First, we describe a user-centered design process for finding functional requirements for holistic control panels supporting better collaboration and coordination between transport operations and supply chain processes affected by or affecting the transport of goods. Secondly, we present the resulting functionality as seen from the perspective of different actors in the supply chain, from producer to shop. One of the largest retailers in Norway is used as a case.

The case study and the user-centered approach are performed with several elicitation methods such as observations, interviews, innovation games and paper prototyping. The suggested solutions are expressed by means of paper prototypes which have been co-created and validated by the stakeholders in the supply chain during an iterative incremental process.

Currently, solely central experts in the organisations involved are able to solve problems by combining information from many sources and by taking the right actions. Due to the identified need for more robust and automated solutions, the paper prototypes suggest unified solutions that (1) provide easy and automated access to the right information at the right time for all actors in the supply chain; (2) supports easy detection of deviations; and (3) supports decisions that can improve efficiency and deviation handling.

Keywords: Retail supply chain · Collaboration · Coordination · User centered design · Transport · Information technology · Paper prototype

1 Introduction

Retail supply chains involve many actors, and they handle large volumes of cargo with short lead times and different requirements with respect to handling and temperature. The actors are typically producers, wholesalers, transport service providers, truck drivers and retailers with different profiles. In addition the wholesalers have central and distributed warehouses and departments working together, assisted by internal information technology (IT) systems. These systems also communicate with external systems to support efficient coordination and communication for purchasing, order management, replenishment, invoicing, etc. The retail supply chain eco system is getting more

© Springer International Publishing AG 2016
G. Fahrnberger et al. (Eds.): I4CS 2016, CCIS 648, pp. 192–208, 2016.
DOI: 10.1007/978-3-319-49466-1_13

advanced as new technology is deployed. ASR (Automatic Store Replenishment) does for example simplify the replenishment process, GPS and temperature sensors are commonly used in trucks to enable continuous tracking of the distribution process, and RFID tags on pallets and other load units facilitate automated registration of shipped and received goods. The solutions also support data collection that can support prognosis and decisions.

IT support to transport operations has to a large extent focused on single actors, e.g. transport management systems used by actors responsible for the transport. [1] addresses the role of IT in collaborative decision making in supply chains and shows positive effects on customer service performance, but states that further research on this issue and dyadic relationships in supply chains are requested, among others between shipping and transport service providers. The use of tracking technology is studied by [2]. Such technology is so far mainly motivated by the needs of the wholesalers. Suppliers have in general adopted tracking technology to strengthen the relationship towards their main customers and not for the sake of their own production process. The importance of an actor perspective is highlighted since there are different needs and motivations among the parties in the supply chain, and more research with such an actor perspective is requested, among others the perspective of transport service providers.

This paper incorporates the perspectives of the different actors in the supply chain, and provides a holistic view upon the transport operations where coordination and collaboration among all parties involved is emphasized. The aim has been to identify requirements for IT support enabling more optimal coordination of tasks and processes directly and indirectly related to transport operations. An iterative approach with user-centered design is carried out in close collaboration with actors in the supply chains of one of the largest wholesalers in the retail sector in Norway. Personnel employed in different positions in the wholesaler's organisation, producers, shops, transport service providers and truck drivers have been involved.

The next sections provide an overview of the method used, the current situation, and the results which are a conceptual view upon a unified collaboration console for transport and associated paper prototypes. The results are discussed from the perspective of different stakeholders and conclusions are made.

2   Method

The coordination between the transport operations and the various operations of the stakeholders in the supply chain is currently done by verbal communication (face-to-face or by phone), communication through documents (such as freight papers) and communication through various IT systems. The aim is to improve this coordination and collaboration by providing information that is not currently easily available to support situational awareness and to assist decision making. Our hypothesis is that better IT tools for information sharing in the transport chain will lead to (1) more efficient access to just in time and just enough information; (2) better handling of deviations; and (3) more informed decisions by all parties involved.

Our proposed solution is to merge information from various channels into holistic views specific for the different stakeholders. The suggested artifact is a design of a system that takes the different needs of the stakeholders into consideration and offers holistic and up-to-date views.

The aim of this work is to answer the research questions: RQ1: What are the information needs of the parties involved in the transport chain in order to enable better handling of and adaption to deviations? RQ2: What are the information needs of the parties involved in the transport chain in order to take more informed decisions?

2.1 Research Method

The design science research method is used according to [3], as depicted in Fig. 1. The research environment consists of all stakeholders involved in the retail supply chain, starting from the producer, to a wholesaler with different departments and warehouses and further to the final selling point, all by means of transportation delivered by transport service providers. The research is driven by the need to better coordinate and share information in this environment, with the assumed outcome a more optimal operation and therewith cost reductions. The knowledge base is based on the existing literature on the use of technology in retail supply chains as well our own findings.

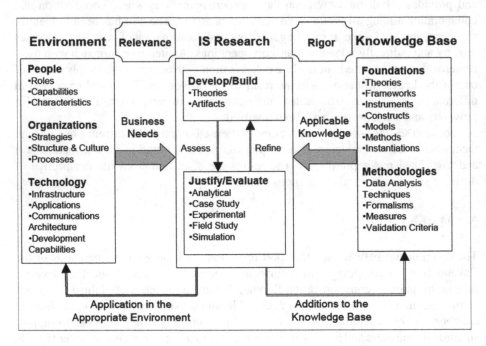

Fig. 1. Conceptual framework for IS research [3]

The information system (IS) research *refine* and *assessment* process, as depicted in Fig. 1, is carried out according to a design methodology inspired by the Lean Startup product design approach [4]. The product we build is however a paper prototype artefact, not actual software. The detailed approach is illustrated by Fig. 2 and focuses on quick cycles through three distinct activities (learn, build and measure). Each activity aims at creating a result (ideas, products and data) where the fidelity of the results improves gradually towards an artefact as we iterate through the whole process.

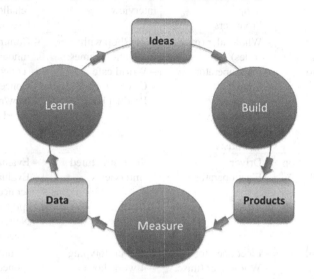

Fig. 2. The Lean Startup [4]

The first *ideas* were based on initial studies of the current situation, and the *build* activity designed a set of products, in our case paper prototypes of software solutions. These paper prototypes were evaluated (*measure*) in evaluation activities, resulting in *data* about the validity of the initial hypothesis. During the *learn* activity, data are analysed, new insights and new *ideas* gained that is input to a new build activity.

2.2 User-Centered Iterative Design

User-centered design is conducted prior to the development of complex systems to ensure deep understanding of user and stakeholder roles. The aim is to ensure that system designed support the daily work of end users and the role of stakeholders [5, 6].

User-centered design is applied in all activities in the iterative cyclic process described above. The activities are carried out in close cooperation with real stakeholders by means of various methods for data collection, as described in the sections below.

2.3 Data Collection

Table 1 summarises the steps taken to collect data on user needs and to elaborate the solution, which stakeholders have been involved, which methods have been used for data collection and what the results were.

Table 1. Steps taken to collect data

Steps taken	Stakeholders involved	Method	Results
Learn – 1st iteration: As is situation mapping	– Wholesaler (misc. roles) – Fleet operator – Shops – Producers	Case study – Observation – Semi structured interviews	– As-is situation mapping – Awareness of challenges – User needs
Build: Transport collaboration console concept definition	– Wholesaler (misc. roles) – Fleet operator – Shop – Producers – Technology provider – Scientists	Workshop with innovation games: – World cafe – Cover story – Product box	– Common understanding of concept – Concept definition – New/refined user needs requirements
Measure: Evaluation of previous trials	– Driver – Fleet operator	– Semi structured interviews	– Evaluation of trials – Evaluation of concept/paper prototype – New/refined user needs requirement
Measure, learn, build: Evaluation, reflection and further elaboration of the transport collaboration console	– Fleet operator – Wholesaler (misc. roles) – Technology provider	Paper prototyping – 4 workshops	– Evaluation of concept/paper prototypes – Improvement suggestions – New/refined user needs/requirements – New/refined paper prototypes adapted to user needs and requirements
	– The above – Shop – Producers	– World cafe – Semi structured interviews	

The *as is* situation was mapped by a **case study** where semi-structured interviews and observations were used to collect data. This provided an overview of the current situation and the specific needs and challenges of the different stakeholders. We catered for an open dialogue to understand user expectation from the sought after solution. Questions therefore primarily start with how, what and why, e.g. "How do you do your work today?" followed up with trigger questions such as "what irritates you during a working day?"

To define the overall concept of and expectations for a transport collaboration console we played **innovation games** [7] such as world café, cover story and product box to gather input from the stakeholders involved in the transport chain. The *World Café* [8], game was used to identify challenges faced in the transport chain. This game facilitates a structured conversational process with an open discussion. Groups of participants, representing different stakeholder groups, visited the "café table" where they continued to discuss partly based on the results from the previous group.

After the *World Café* we used the *Cover Story* game to identify and elaborate on the long term vision of the work on a transport collaboration console. The participants wrote the headlines that they would like to see in the newspaper after a successful implementation of the console.

In the *Product Box* game the participants were asked to present and "sell" the transport collaboration console as a product by creating a physical cardboard box with printed slogans about the features of the product. In this way, the main features that are of importance for the participants get highlighted, which again gives insight in the desired outcome.

We also carried out evaluations of trials carried out by the wholesaler in collaboration with a transport services provider and shops such as use of RFID to support automatic registration of pallets received by the warehouse and automatic registration of pallets in trucks; and also notifications of shops 30 min in advance of the arrival of the truck. The data collection was carried out by means of semi-structured interview.

The elaboration of the actual functionality provided by the transport collaboration console was facilitated by **paper prototyping** [9] together with the different stakeholders. Layout and visual effects were not emphasized, just the functionality and the information to be presented and shared. This was an iterative process starting with initial prototypes based on the above. We also used the *World Café* and semi-structured interviews process at a later stage to verify and to get further input on the paper prototypes from stakeholders other than those participating in the initial work.

3   Case Study: As-Is Situation

A case study was carried out to capture information about the current processes and information flows. The overall findings are depicted in the BPML (Business Process Modeling Language) diagram. The diagram for outgoing transport is depicted in Fig. 3. Each actor involved is represented by a pool which contains the processes involved. The processes directly involved in the transport tasks are green, and information flows (black dotted lines) and status information (red dotted lines) to and from these processes are shown. The others processes show the context in which the transport tasks are executed. The solid arrows are control flows between processes.

The actors directly and indirectly involved in transport are the retailer, the transport service providers, the truck drivers, the wholesaler's local distribution centre, the wholesaler's central warehouse and producers delivering products to the wholesaler. The focal point is the wholesaler's distribution centre which is represented by two of its main units – the customer service centre and the warehouse.

3.1   Outbound Transport

Outbound transport from the wholesaler's local distribution centre to the shops is based on orders coming from the retailers. Information on the volumes to be provided to each shop is transferred to the transport service provider to initiate the transport planning, which use route plan templates optimized based on prognosis as the starting point. The

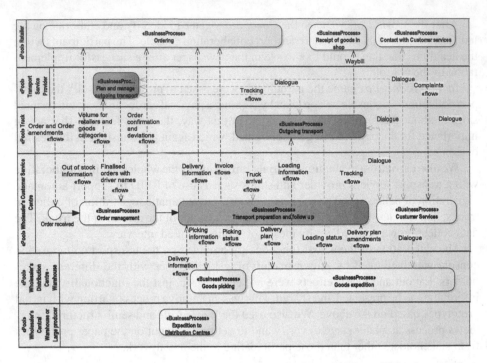

Fig. 3. As is situation - outgoing transport processes and information flows (Color figure online)

route plan templates (previously received from the wholesaler) are adapted to the actual volumes ordered by each shop, and finalized orders assigned to drivers, routes and trucks are generated. The shops will get access to information which confirms or disapproves (e.g. in case of out of stock) the delivery of the products listed in the orders.

The transport will be controlled by the wholesaler's transport preparation and follow up process. On arrival to the warehouse, the trucks will be assigned a loading gate. Outgoing transport is composed of (1) pallets picked from the warehouse at the local distribution centre and (2) pallets delivered for cross-docking from the wholesaler's central warehouse or from large producers of fresh food. Information about the latter is received from the shippers. The first is established by the picking process of the local warehouse according to received picking information. The picking is monitored so that the transport preparation and follow up process is aware of the status with respect to the picking for each shop and each outgoing truck.

The goods expedition process in the warehouse has for each truck access to information about all the pallets to be delivered to each shop and the associated temperature requirements. This includes pallets picked at the local warehouse as well as cross-docked pallets, and the loading information is provided to the respective truck drivers, who are responsible for the loading. The loading status is monitored. It may happen that pallets cannot be loaded due to space limitations. In such cases the customer service centre is informed, and the customer must approve that the delivery of some of the pallets is postponed. If there is space available in the truck, the shop will be asked to approve that the wholesaler pushes goods to fill up the truck. In case of changes (postponed deliveries or pushed goods), the delivery plan is amended, and the final delivery information is made available to the retailer. An invoice will also be sent to the retailer based on this information.

The truck is tracked during the transport, and the transport service provider can follow the progress. On arrival to the shop, the truck driver will present the waybill to the retailer and collect a signature.

Several challenges were discovered during the work on the process description. The trucks may arrive to the shops in a 3 h time slot, and shops do not know the exact arrival time in advance. Similarly, the local distribution centre do not get the exact arrival time of inbound trucks, which for example may carrying pallets for cross-docking. Most of the coordination, status reporting and deviations are handled manually. This means that actors communicate by phone. Usually this works well, but it is resource demanding, and in some cases the procedures fails. The shops may for example not get information about delays and cargo that cannot be delivered due to space limitations in the vehicles. It also takes time to detect that pallets are delivered to other shops than planned.

3.2 Incoming and Intermediate Transport

The wholesaler orders deliveries from producers and the deliveries are produced and delivered according to agreed time slots. The inbound transport to the wholesaler is in this study booked by the producers to facilitate the best possible coordination with the production process. However, all transport service providers used have contracts with the wholesaler. As for the shops in the study of the outbound transport, the producers do not know the exact time of arrival for the trucks.

The inbound transport is whenever possible organized as return load. The trucks used on outbound transport will when they return pick up deliveries from producers on their way back to the wholesaler's warehouses. The warehouses expect the deliveries within a time frame, but do not know the exact time of arrival in advance.

The warehouse of the wholesaler's local distributions centre will in addition to inbound transport from producers also get deliveries from the wholesaler's central warehouse. The latter may be pallets for cross-docking as well as deliveries that are to be stored. As in the above case, they do not know the exact time of arrival in advance.

3.3 Existing IT Solutions

The wholesaler and the shops use advanced IT systems comprising functionality such as forecasting and replenishment, warehouse management, vendor managed inventory and order management. It is possible to get access to information about almost everything if you know how to use the system. The wholesaler also has expert users who look up and manually combine information from many different sources to get an overview over situations and to manually take decision on how to handle deviations. It is however quite common for other users to ask for information via manual communication.

The wholesaler provides information to transport service providers, producers and retailers via dedicated portals. The information is however not complete, and they frequently contact each other by phone. The producers and retailers also communicate with the transport service providers and their truck drivers to handle deviations and to coordinate their activities.

The trucks are equipped with temperature sensors and GPS tracking equipment so the transport service providers are able to locate the trucks and check the temperature for the different temperature zones in real time. Previous trials have used GPS tracking to implemented SMS notifications to retailers 30 min in advance of the actual arrival of the trucks. Previous trials have also tested the use of RFID in inbound and outbound transport. Pallets equipped with RFID and RFID readers in warehouse gates and in trucks enable automated registration of receipts of goods in warehouse and goods loaded onto trucks.

4 Results

This section summarizes the results from the user-centered design approach which include a definition of a unified transport collaboration concept and paper prototypes.

4.1 Unified Transport Control Concept

The case study and the innovation games carried out with the stakeholders in the initial phase of the work provided input to a conceptual view upon the desired result – i.e. control console for transport related tasks providing (1) a holistic view on the transport operations; (2) support for information access related to all types of transport; (3) support for coordination between actors and processes involved in and affected by the transport; and (4) decision support in case of deviations.

Figure 4 provides an overview of the concept. The different stakeholder types will have access to the desired functions via tailor-made views. When realised, these views will be user interfaces adapted to the needs of each stakeholder type. The views present information to support the users in a way that fulfils their needs. The wholesaler already has advanced legacy systems with relevant data sources and data elements. In addition there may be other data sources, e.g. real-time tracking and temperature data from trucks. The functions will access, combine and process the data to provide the required functionality to the stakeholders.

User needs with respect to the transport collaboration console were collected through interviews, evaluations of trials and also during the paper prototyping as listed in Table 1. Several needs addressed easy access to information in an overview picture or just a click away from such a picture. The use of a map with clickable objects representing warehouses, shops and routes as well as the real-time position of trucks were suggested. Other needs concerned notification about statuses and deviations as well as functionality. An important observation was that many of the different stakeholders requested the same information or functionality, which confirmed the overall concept depicted in Fig. 4. In an iterative process, requirements to the information sharing and functionality were derived from the user needs and expressed by means of paper prototypes.

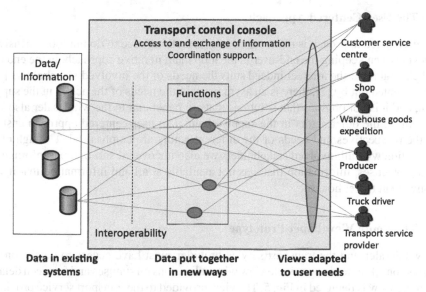

Fig. 4. The transport collaboration console concept

4.2 Paper Prototypes

The iterative approach has led to the establishment of several paper prototypes. As described above, layout and visual effects are not emphasized. The paper prototypes address information and functionality needed by different stakeholders in different situations by means of a set of views as listed in Table 2.

Table 2. Views elaborated in paper prototypes

View	Description
Map	Provides overview with clickable routes, trucks, shops, warehouses
Routes	Provides statuses for all routes/ trucks (foreseen delays, available capacities, etc.)
Truck – in/out	Supports follow up of transport operations by providing truck status (foreseen arrivals, delays, faulty deliveries, temperature, etc.)
Warehouse – in	Supports cargo reception and cross-docking
Warehouse – out	Supports goods expeditions from warehouse
Shop	Supports goods reception in shops
Producer	Supports the producer's goods expedition

5   Discussion and Lessons Learned

The user-centered approach and a selection of the resulting paper prototypes are discussed below.

5.1   The User-Centered Approach

A user-centered approach is often viewed as 'something nice to do but too expensive', however, when keeping the end-user in the loop in an iterative approach, in the end one can be certain that the artefact indeed suits the needs of the involved stakeholders [5].

As pointed out by [2], there is an asymmetry in the needs of the actors in the supply chain, and it is not obvious that solutions that are beneficial to the wholesaler also will be useful to the other actors in the supply chain. The user-centered approach ensures that the perspectives and needs of the different actors are emphasised. Through close cooperation with all involved stakeholders we also uncovered sub-optimal information sharing practises, information that was not available at all and information that it was inconvenient to get hold of.

5.2   The Route View Paper Prototype

The wholesaler and the transport service provider must have access to details on the progress on all routes in one single view with indications on statuses and foreseen delays. The route view is depicted in Fig. 5. The view provided to the transport service providers will be limited to the routes they operate. For each truck it provides the registration number, planned and actual departure from warehouse and a list of shops with planned, expected and actual arrivals and deviation. It also provides the number of pallets on board for each cargo category (dry, cold, frozen and fruit & vegetables (F&V)).

Routes from warehouse XYZ
Current time 09:10

Truck	Planned dep.	Actual dep.	Shop	Planned arrival	Expected arrival	Actual arrival	Devia-tion	Dry	Cold	Frozen	F&V	Sum	Return	Free space (pallets)
										# pallets				
VH12345	05:30	06:15						50	4	3	2	59	30	7
			Shop A	06:00-08:00	07:05	07:06		20	2	1	1	24	30	1
			Shop B	10:15-12:15	11:20			10	2	2	0	14		15
			Shop C	16:30-18:30	17:45			20	0	0	1	21		36
VH23456	06:00	07:00						13	4	7	7	31		35
			Shop A	08:00-10:00	10:00			1	1	1	1	4		39
			Shop D	10:15-12:15				2	2	4	0,5	8,5		47,5
			Shop E	16:30-18:30				10	1	2	5,5	18,5		66
VH34567	06:00	08:00						6	15	24	15	60		6
			Shop X	07:00-09:00	09:30			1	4	7	6	18		24
			Shop Y	10:15-12:15				2	5	8	5	20		44
			Shop Z	16:30-18:30				3	6	9	4	22		66
											Delayed/deviation			
											May be delayed			
											As planned/delivered			

Fig. 5. Overview of routes

Truck should always be as full as possible (within the volume and weight limits imposed) in order to use the resources in an optimal manner and to contribute to fewer vehicles on the road and reduced fuel consumption. To address such issues, the view supports information exchange and decisions on space utilisation. The amount of return load from each shop can be entered, and the total availability of empty space after the visit to each shop is presented to support the planning of return load.

5.3   The Truck In/Out View Paper Prototypes

The wholesaler and the transport service provider must be able to follow the progress of individual trucks. The information needed for incoming, intermediate and outgoing trucks are more or less the same, so we are just showing one of the views in Fig. 6.

Current time 14:10				Type of product		Dry	Chilled	Frozen	F&V	
				Measured temperature		N/A	4C	-23C	4C	
Faulty deliveries (yes/no): No										
					# pallets per product type					Click for m3
Route	Plan	Arrival	Departure	Deviation	Dry	Chilled	Frozen	F&V	Sum	Rest
Local warehouseTrondheim	11:00		11:20	0:20	4	5	4	2	15	9,5
- Shop B	11:30-13:30	13:15	14:00		2	2	1	0,5	5,5	0
- Shop C	12:00-14:00	14:30		0:30	0	2	2	0	4	4
Checkpoint K	12:30-14:30	15:00								
- Shop A	13:00-15:00	15:30			1	0	0	0	1	1
- Shop F	16:00-18:00				1	1	1	1	4	4
- Shop Y	17:00-19:00				0	0	0	0,5	0,5	0,5

Driver	Tine Maarudsson	Status (driving/stoped): Driving		
Phone	9912323			
Registration number:	VH 12345			
No of pallets - 1st hight	33			According to plan
No of pallets - 2nd hight	66			Delayed

Fig. 6.   Overview of outgoing trucks

The locations to be visited on the route are listed. This may be warehouses, shops, producers and waypoints. For each location the planned, expected and actual arrival and the actual departure are provided. If relevant, deviations are also included. The number of pallets in cargo categories (dry, cold, frozen, F&V) is also provided as well as statuses with respect to faulty deliveries and remaining pallets for each l.

5.4   The Warehouse In/Out View Paper Prototypes

The wholesaler needs an overview on incoming truck to each warehouse to be able to plan and prioritize the reception of the cargo. Figure 7 provides a list of all incoming trucks, the planned arrivals to the warehouse, the departure time and location (from), the expected and actual arrival time and also the amount of cargo of different categories, including cargo for cross-docking. By means of data from the central systems of the wholesaler it is possible to identify trucks with goods that is to be transhipped within a short timeframe and prioritize based on this information.

| Local warehous XYZ | | | | | | | | | Click for m3 | | | | |
| Current time: 04:00 | | | | | | | | | # pallets | | | | |
Planed arrival	Depar-ture	Devia-tion	Expected arrival	Actual arrival	Truck	From	From location	Cross-dock	Dry	Cold	Frozen	F&V	Total
00:00-06:00	23:00		05:00		DL12345	Central warehouse	Oslo	66					66
00:00-06:00	23:24		05:20		DL23456	Central warehouse	Oslo	15	10			30	55
00:00-06:00			N/A	N/A		Pizza producer	Stranda		16	50			66
00:00-06:00			N/A	N/A		F&V wholesaler				10		50	60
00:00-06:00			N/A	N/A		Dairy ABC	Verdal		63				63
06:00-12:00			N/A	N/A		Dairy ABC	Heimdal		20				20
06:00-12:00			N/A	N/A		xxx							0
12:00-18:00			N/A	N/A		Producer BCD							0
						...							

Will be transshiped in less than 2 hours

Fig. 7. Overview of incoming trucks to warehouse

The wholesaler's warehouse goods expedition needs an overview of the status with respect to the picking and loading of pallets for outgoing trucks. Figure 8 provides such an overview of trucks and planned and actual departures and deviations. The sequence of the shops should match the desired loading sequence, and the field in which the cargo for the different shops is placed is also provided. It is easy to follow the status of the picking of goods in each goods category for each truck. In case of delays, decisions that facilitate more efficient picking can be taken. The overview also facilitates that trucks can be assigned gates according to the picking status, and the time the trucks spend by the loading gates can be minimized.

| Local warehouse xyz | | | | | | | | | Click for m3 | | | |
| Current time: 05:30 | | | | | | | | | # pallets | | | |
Planned dep.	Actual dep.	Deviation	Truck	Shop	Field	Cross-dock	Dry	Cold	Frozen	F&V	Total
06:00			VH12345			0	22	11	12	6	51
				- Shop B	2		2	2	1	0,5	5,5
				- Shop C	2		3	8	6	0	17
				- Shop A	3		7	0	0	0	7
				- Shop F	3		10	1	5	5	21
				- Shop Y	3		0	0	0	0,5	0,5
06:30			VH2345			8	20	10	10	7	55
				- Shop Z		8	20	10	10	7	55
										Picked	
										Loaded	

Fig. 8. Overview of outgoing trucks from warehouse

5.5   The Shop View Paper Prototype

When working with the view for the shops we discovered that in some cases an information demand from one part of the supply chain can cause dissatisfaction in other parts. In this case, the shop wanted to know the exact location of the truck which is going to deliver the goods, though the drivers consider this a breach of privacy. The provision of such information to the shop may also cause misinterpretations since the shop neither know the driving and traffic conditions nor the status with respect to the obliged resting hours of the driver. Also a GPS location close to the shop might be misleading as the shop does not know the delivery route, and the shop may not be the first stop on the route. In order to cater for the needs of both parties, we agreed on a notification with the expected arrival time 30-minute in advance, which is confirmed by the driver before it is sent to the shop. This maintains the drivers' privacy and the shop manager has sufficient time to prepare the reception area.

The shop's view in Fig. 9 supports the coordination of deliveries between the transport service provider and the shop, and also the deliveries of extra goods pushed by the wholesaler to the shop based on foreseen campaigns. In case of the latter, the shop may accept or refuse the extra goods. The number of pallets to be provided within each goods category will be presented to support the planning of the cargo reception, space allocation, etc.

Shop XYZ												
Current time:	Monday 1/2		10:30	Arrival				# pallets				
Date	Plan	From	Truck	Expec-ted	Act-ual	Devi-ation	Order	Dry	Cold	Frozen	F&V	Sum
Monday 1/2	10:30-12:30	Warehouse A	VH12345	11:00			12345	1	1	1	1	5
							12346	1				
							22222	1	1	1		3
Wednesday 3/2	14:00-16:00	Dairy ABC	VH23456				56789	0	1	0	0	1
Friday 5/2	07:00-09-00	Warehouse A					23456	2	2	1	2	7
Address	xxx						Extra goods offered by wholesaler					
Phone	xxx						Extra goods approved by shop					
Contact	xxx						Extra goods refused by shop					
Opening hours	08-23(8-21)											

Fig. 9. Overview of incoming trucks to shop

5.6 The Producer View Paper Prototype

The producer's view in Fig. 10 supports the coordination between the producer and the transport service provider. One truck may pick up pallets linked to one or more orders. For each truck the view will present the planned time slots for pick-ups and the associated number of pallets. Planned pick-ups after the normal opening hours are highlighted since they will require special preparations. 30 min ahead of the actual arrivals, the expected arrival time will be announced.

The producer may request deviations from the suggested plan. They can indicate when the goods can be ready for pickup and suggest changes in the number of pallets if the requested quantum cannot be produces or if more goods can be provided. In case of the latter, information on the free capacity in the trucks is used when the number of extra pallets is suggested. The suggested changes must however be accepted by the transport service provider.

Monday 1/2				Current time: 10:30							
				Arrival			Act-ual dep	Ready	# pallets		
			Order no.	Planned	Expec-ted	Divi-ation			Order	Chan-ge	Free space
Date	Receiver	Truck									
Monday 1/2	Warehouse A	VH12345	12340 / 12345	10:30-12:30	11:00			1/2 / 11:00	4 / 8	0 / -4	6
Monday 1/2	Warehouse A	VH2346	12345 / 12347	12:30-14:30				1/2 / 12:30	4 / 4	+4 / +3	8
Thursday 4/2	Warehouse B	VH12345	23447 / 23456	17:00-19:00				4/2 / 15:00	5 / 10	-3 / +5	5
Friday 5/2	Warehouse C	VH12345	23456	07:00-09-00				4/2 / 16:30	10	-5	10

Producer xxxxx		Accepted
Address xxxxx		Resused
Phone	45 519 272	Not accepted/refused
Contact person	NN	Colunm for input
Opening hours	08:00-16:30	Time earlier/later

Fig. 10. Overview of pick-ups at producer's location

6 Conclusion

The asymmetric needs of the actors in the retail supply chain are a challenge when new technology is to be introduced. The actor perspective and user-centered approach used in the work described by this paper is one of the measures that have ensured that the views of all actors involved have been considered. These views are defined in paper prototypes. Due to the identified need for better collaboration and coordination support, the paper prototypes show unified solutions that (1) provide easy and automated access to the right information at the right time for all actors in the supply chain; (2) supports easy detection of deviations; and (3) supports decisions that can improve efficiency and deviation handling.

The involvement of the relevant actors in the user-centered design has provided answers to our research questions. RQ1 and RQ2 are answered by the paper prototypes. They define the information needs of the parties involved in the transport chain in order to enable better handling of and adaption to deviations and with that information they are enabled to take better informed decisions.

6.1 IT Solutions

The case study showed that the IT solutions required to enable the desired functionality are partly in use or partly in the stage of being adopted. The data collection from GPS sensors in transport means is crucial for situational awareness and detection. By comparing positions with distribution plans and time schedules it is possible to detect delays and to announce arrivals to shops and producers. Temperature sensors support detection of deviations in cargo condition during transport and transshipments. Cargo units labelled with RFID tags and RFID readers in warehouses, shops and transport means arrange for automated and reliable tracking of cargo units and thus also detection of deviations.

The internal IT system of the wholesaler is the nave in the coordination of cargo handling processes, but currently the dedicated portals provided to transport service providers, shops and produces do not include sufficient information on statuses and deviations. The systems must be extended to support the views defined by the paper prototypes. Events and deviations can be detected or predicted based data collection from the sensors and RFID readers described above.

The views defined by the paper prototypes can be realised as extensions of the IT systems of the wholesaler, but may also be independent applications, for example apps running on mobile phones. Such applications must have open interfaces for information exchange with suppliers and customers. The integration with the IT systems of the wholesaler should also be implemented via these interfaces.

6.2 Decision Support

The current IT systems provide useful information and functionality, but there is a need for expert users who know how to find and combine the relevant information elements in order to establish an overview that supports decisions. Other users are to a large extent informed by the expert users via manual communication (mainly phone). More easy access to information and better decision support can make the expert user even more efficient and also lower the pressure on these users by supporting other users.

The functionality defined by the paper prototypes supports situational awareness and detection of deviations. The outgoing truck view will for example indicate foreseen deviations to inform shops and to support decisions on how to handle the situation. The view also arranges for coordination. The producer may for example indicate when products are ready for pick-up and the quantity that will be ready.

Better situational awareness and information on deviation arrange for better self-coordination. Shop managers and producers who are notified about the time for arrival of the truck can schedule preparations and do resource allocations that are better adapted other activities. The wholesaler warehouse can use information on foreseen arrival of trucks when the use of resources is planned, and trucks with goods for cross docking can be prioritized. The goods expedition can allocate gates to trucks according to picking updated statuses.

6.3 Further Work

It remains to implement and test the functionality defined by the paper prototypes and to evaluate the effects for the different actors in the retail supply chain. According to [2] it is important that the wholesaler lowers the implementation barriers of other actors in the supply chain since the wholesaler is the actor with the most power and also the actor who probably will benefit most from better IT solutions. Hinkka states that this can be done by alignment of roadmaps; by understanding the challenges of the other actors; by taking the main responsibility for the implementation; by using the wholesaler's negotiation power in discussions with system providers; and by purchasing equipment for the other actors to get discounts. In the case addressed by this paper the wholesaler controls most of the information as well at the systems that can control the information flows in the supply chain. Thus, the wholesaler should take the leading role and implement IT support for the functionality required by the other stakeholders. The paper prototypes define the requirements for such an implementation.

Acknowledgments. The authors gratefully acknowledge the Norwegian Research Council and the BIA program for financial support of the project as well as the participating case companies, which together enabled this study.

References

1. Wong, C.W.Y., et al.: The role of IT-enabled collaborative decision making in inter-organizational information integration to improve customer service performance. Int. J. Prod. Econ. **159**, 56–65 (2015)
2. Hinkka, V., Främling, K., Tätilä, J.: Supply chain tracking: aligning buyer and supplier incentives. Ind. Manage. Data Syst. **113**(8), 1133–1148 (2013)
3. Hevner, A.R., March, S.T., Park, J., Ram, S.: Design science in information systems research. MIS Q. **28**(1), 75–105 (2004)
4. Ries, E.: The Lean Startup: How Today's Entrepreneurs Use Continuous Innovation to Create Radically Successful Businesses. Crown Business, New York (2011)
5. Kubie, J., Melkus, L.A., Johnson, R.C., Flanagan, G.A.: User-centred design. In: Brown, C.V., Topi, H. (eds.) IS Management Handbook, 7th edn. CRC Press, Florida (2000)
6. Shluzax, L.A., Steinert, M., Katila, R.: User-centred innovation for the design and development of complex products and systems. In: Plattner, H., Meinel, C., Leifer, L. (eds.) Design Thinking Research: Understanding Innovation, pp. 135–149. Springer, Cham (2014)
7. Gray, D., Brown, S., Macanufo, J.: Gamestorming: A Playbook for Innovators, Rulebreakers, and Changemakers. O'Reilly Media, Sebastopol (2010)
8. Brown, J., Isaacs, D., Wheatley, M.J.: The World Café: Shaping Our Futures Through Conversations That Matter. Berrett-Koehler Publishers, San Francisco (2005). ISBN 1576752585
9. Rettig, M.: Prototyping for tiny fingers. Commun. ACM **37**(4), 21–27 (1994). doi:10.1145/175276.175288

Short Papers

Potentials and Requirements of an Integrated Solution for a Connected Car

Karl-Heinz Lüke[1], Gerald Eichler[2], and Christian Erfurth[3(✉)]

[1] Ostfalia University of Applied Siences, Wolfsburg, Germany
ka.lueke@ostfalia.de
[2] Telekom Innovation Laboratories, Darmstadt, Germany
gerald.eichler@telekom.de
[3] Ernst Abbe University of Applied Siences, Jena, Germany
christian.erfurth@eah-jena.de

Abstract. Within our society, individuality, comfort and mobility play an important role for working, travelling and living environments. Especially with the connectivity of a car to its environment, e.g. sensors, the communication to another car, to infrastructure communication and provisioning of relevant information to driver, can be considered as valuable technology of the future. An integrated solution for a connected car can be used for a safe, intelligent and comfortable mobility. Although there are significant results from research projects concerning car-to-x, e.g. sim^{TD}, and other infotainment and entertainment projects available, an integrated concept that covers general technical requirements, the drivers' needs and business aspects is hard to find on the market. Significant requirements derived from technical and market insights are evaluated. These findings reflect the introduction of an integrated architecture that covers car-to-x communication, info- and entertainment and IT-security aspects.

Keywords: Car-to-x · Integrated architecture · Communication infrastructure · Technical requirements · Market insights

1 Current Situation and Motivation

Cars are considered as autonomous systems that have their own infrastructure. In many cases driving assistant systems, relying on CAN bus, the navigation system, based on GPS and TMC and infotainment systems with radio, music, TV or gaming options as well as communication to mobile networks are still working independently. Looking at a scenario from the users' point of view, Fig. 1 shows a typical week day, assuming the user relies on personal electro-mobility for travelling between home and work. Re-active handling needs to be transferred into pro-active guidance for information providing to the user's needs.

The key device for information and communication remains the personal smartphone of the driver, which is often low compatible with the car control and infotainment system. Seamless smartphone integration can enhance the machine-to-driver interface (M2D) in terms of comfort and driving pleasure.

© Springer International Publishing AG 2016
G. Fahrnberger et al. (Eds.): I4CS 2016, CCIS 648, pp. 211–216, 2016.
DOI: 10.1007/978-3-319-49466-1_14

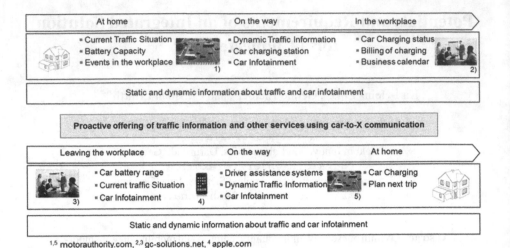

Static and dynamic information about traffic and car infotainment

Proactive offering of traffic information and other services using car-to-X communication

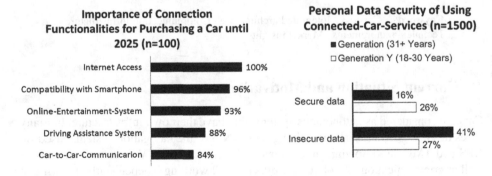

Static and dynamic information about traffic and car infotainment

1,5 motorauthority.com, 2,3 gc-solutions.net, 4 apple.com

Fig. 1. Information assisted week day scenario from users' point of view

A recent study in Germany (Fig. 2a) pointed out that Internet access (100 %) is the most significant feature of connection functions for purchasing a car. Compatibility with smartphone (96 %) and online entertainment (93 %) becomes essential for more and more people. Currently, driving assistance systems (88 %) and car-to-x communication (84 %) are essential but seems not important as other functionalities. Respectively, with the growing diffusion rate of these features in the near future, they will become very relevant by purchasing a new car.

Fig. 2. (a) Car connection functions, market survey (n = 100) [11]; (b) Data security, market survey (n = 1500) [12]

As soon as integrated systems are discussed, the key question "How secure are personal data of using connected car services?" arises. The Generation Y (18–30 years) considered to be the generation of digital natives and is much more familiar with personal data provisioning to connected services. In contrast, the older generation (>30 years) is more restrictive when providing personal data to Web services, see Fig. 2b. Consequently, data is important and should be considered for an integrated solution.

2 Integrated Solution for Car-to-X Communication

The term "Car-to-X" covers a wide spectrum of interaction targets (Fig. 3), where the car serves the role of central intelligence. Probably, the car-to-car aspect covers the most interesting area, as multiple cars form a Mobile Ad-hoc Network (MANET). As most modern cars carrying their own SIM card, the master-slave-relation between the car communication system and user smart phone needs so be resolved smoothly.

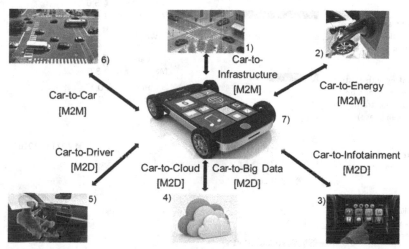

1,6 motorauthority.com, 2 focus.de, 3 pcworld.com, 4 cloudlinktech.com, 5 blog.caranddriver.com, 7 magisteradvisors.com

Fig. 3. Car-to-X source and target systems

On the other hand, a car-to-car MANET can be deployed as a multi-sensor meshed network, offering completely new features regarding traffic control strategies, energy savings and autonomous driving. In general, it can be distinguished between machine to machine (M2M) and machine to driver (M2D) relationships. Whereas the M2M (e.g. car-to-car) interface is examined in many projects, M2D (e.g. car-to-driver and car-to-entertainment) will become much more important [1–3].

3 Requirements Derived from Market Insights

For an integrated solution, market insights and technology drive should be considered for the development. A recent study in Germany showed the following trends seen from market (see Fig. 4(a)):

- Better enrichment of navigation information with real time information
- Fast transmission of information in case of an accident

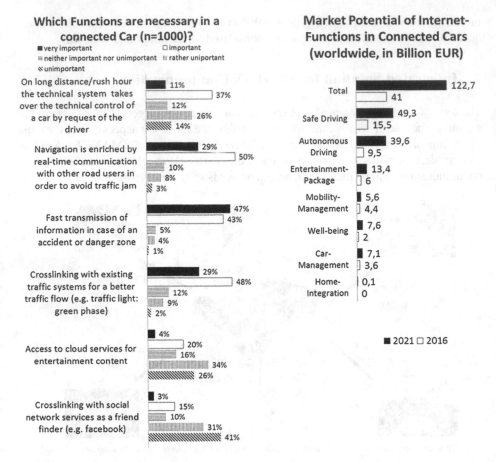

Fig. 4. (a) Necessary functions, market survey (n = 1000) [10]; (b) Market potential of connected car functions till 2021 [9]

- Improvement of a better traffic flow
- Autonomous driving functions in special traffic situations, e.g. rush hour
- Crosslinking with cloud and to social networks are still not in the in-car-focus

Market studies take especially care of the potential revenues worldwide, reachable by Internet functions resulting of both, machine to driver (M2D) and machine to machine (M2M) interaction. A factor of three in total for revenue growth is seen for the next five years (Fig. 4(b)). Especially safe driving, autonomous driving and entertainment functions are estimated to have the highest market potential.

4 Requirements Derived from Technical Aspects

The technical aspects can be bundled into several groups, according to the variety of the x value (see Fig. 5). Having an extension of established car bus systems in mind, security

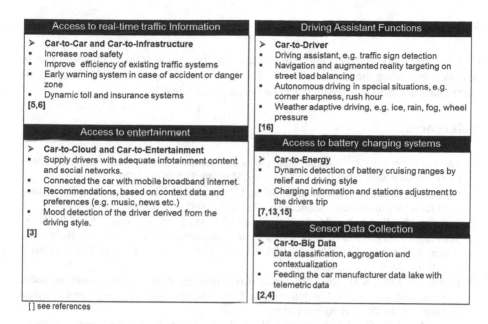

Access to real-time traffic Information	Driving Assistant Functions
➤ **Car-to-Car and Car-to-Infrastructure** ▪ Increase road safety ▪ Improve efficiency of existing traffic systems ▪ Early warning system in case of accident or danger zone ▪ Dynamic toll and insurance systems [5,6]	➤ **Car-to-Driver** ▪ Driving assistant, e.g. traffic sign detection ▪ Navigation and augmented reality targeting on street load balancing ▪ Autonomous driving in special situations, e.g. corner sharpness, rush hour ▪ Weather adaptive driving, e.g. ice, rain, fog, wheel pressure [16]
Access to entertainment	**Access to battery charging systems**
➤ **Car-to-Cloud and Car-to-Entertainment** ▪ Supply drivers with adequate infotainment content and social networks. ▪ Connected the car with mobile broadband internet. ▪ Recommendations, based on context data and preferences (e.g. music, news etc.) ▪ Mood detection of the driver derived from the driving style. [3]	➤ **Car-to-Energy** ▪ Dynamic detection of battery cruising ranges by relief and driving style ▪ Charging information and stations adjustment to the drivers trip [7,13,15]
	Sensor Data Collection
	➤ **Car-to-Big Data** ▪ Data classification, aggregation and contextualization ▪ Feeding the car manufacturer data lake with telemetric data [2,4]

[ ] see references

Fig. 5. General technical requirements for an integrated solution

and persistence layers, as known from airplanes, need to be added as well as wireless communication features with QoS to be improved.

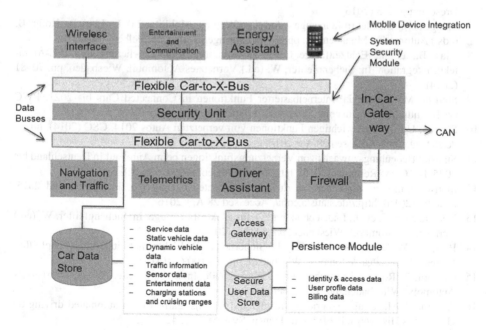

Fig. 6. Revised basic architecture for car-to-x

5 Architectural Concept

Bringing the market and technology aspects together, a car-to-x architecture can be designed, extending the sim^{TD} testbed [1, 8, 14] and improving the car CAN bus (Fig. 6). Stored data needs to be classified according the privacy level.

References

1. Stübing, H., et al.: Sim^{TD}: A Car-to-X system architecture for field operational tests [Topics in Automotive Networking]. IEEE Commun. Mag. **48**(5), 148–154 (2010)
2. Häberle, T., Charissis, L., Fehling, C., Nahm, J., Leymann, F.: The connected car in the cloud: a platform for prototyping telematics services. IEEE Softw. **32**(6), 11–17 (2015)
3. Baltrunas, L., Kaminskas, M., Ludwig, B., Moling, O., Ricci, F., Aydin, A., Lüke, K.-H., Schwaiger, R.: InCarMusic: context-aware music recommendations in a car. In: Huemer, C., Setzer, T. (eds.) EC-Web 2011. LNBIP, vol. 85, pp. 89–100. Springer, Heidelberg (2011). doi:10.1007/978-3-642-23014-1_8
4. Eichler, G., Lüke, K.H., Reufenheuser, B.: Context information as enhancement for mobile solutions and services. In: 13th International Conference on Intelligence in Next Generation Networks, ICIN 2009, Bordeaux (2009)
5. Fuchs, H., et al.: Car-2-X. In: Winner, H., et al. (eds.) Handbuch Fahrerassistenzsysteme - Grundlagen - Komponenten und Systeme für aktive Sicherheit und Komfort, 3, pp. 525–539. Auflage, Wiesbaden (2015)
6. Schäfer, J., Klein, D.: Implementing situation awareness for Car-to-X applications using domain specific languages. In: 77th IEEE Vehicular Technology Conference (VTC Spring), Dresden, pp. 1–5 (2013)
7. Spiegelberg, G.: Elektrofahrzeuge – Auf dem Weg zur Mobilität 2.0. In: Ebel, B., Hofer, B. (eds.) Automotive Management, pp. 57–80. Springer Verlag, Heidelberg (2014)
8. Glas, B., et al.: Echtzeitfähige Car-to-X Kommunikationsabsicherung und E/E-Architekturintegration. In: Siebenpfeiffer, W. (ed.) Vernetztes Automobil, Wiesbaden, pp. 70–81 (2014)
9. Statista: Marktpotenzial internetbasierter Funktionen in Connected Cars bis 2021. PWC (2016). http://de.statista.com/. Accessed 29 Apr 2016
10. Statista: Umfrage zu wichtigen Funktionen von vernetzten Autos 2014. CSC (2014). http://de.statista.com/. Accessed 28 Apr 2016
11. Statista: Bedeutungszuwachs von Vernetzungsfunktionen beim Autokauf in Deutschland bis 2025. BITCOM Research (2016). http://de.statista.com/. Accessed 28 Apr 2016
12. Statista: Umfrage zur Datensicherheit bei connected car services in Deutschland 2015. Deloitte (2016). http://de.statista.com/. Accessed 28 Apr 2016
13. Hechtfischer, K. et al.: Ladetechnik und IT für Elektrofahrzeuge. In: Siebenpfeiffer, W. (ed.) Vernetztes Automobil, Wiesbaden, pp. 82–88, (2014)
14. Paar, Ch., Wolf, M., von Maurich, I.: IT-Sicherheit in der Elektromobilität. In: Siebenpfeiffer, W. (ed.) Vernetztes Automobil, Wiesbaden, pp. 95–100 (2014)
15. Conradi, P.: Reichweitenprognose für Elektromobile. In: Siebenpfeiffer, W. (ed.) Vernetztes Automobil, Wiesbaden, pp. 179–184 (2014)
16. Aeberhard, M., et al.: Experience, results and lessons learned from automated driving on Germany's Highways. IEEE Intell. Transp. Syst. Mag. **7**, 42–57 (2015)

ICT-Systems for Electric Vehicles Within Simulated and Community Based Environments

Volkmar Schau[✉], Sebastian Apel, Kai Gebhard,
Marianne Mauch, and Wilhelm Rossak

Department of Computer Science, Friedrich-Schiller-University Jena,
Ernst-Abbe-Platz 2, Jena, Germany
{volkmar.schau,sebastian.apel,kai.gebhard,
marianne.mauch,wilhelm.rossak}@uni-jena.de

Abstract. The current living standard of industrial nations causes increasing CO_2 emissions, particulate matter, and noise pollution. An essential amount of these environmental issues is induced by stop-and-go traffic within cities which is seriously characterized by short-distance freight transport trips with inner-city and suburban distances. The project Smart City Logistik (SCL) strives for a practical and short-term solution to this problem by ICT-Systems for electric vehicles (EVs). But planning, monitoring and analyzing for urban area logistics can become complicated and challenging to use. Evaluating them within acceptance tests requires a lot of experiments as well as a lot of equipment. The following approach within the SCL project, funded by the German Federal Ministry for Economic Affairs and Energy (BMWi), tries to use the ICT-system as it is and connects that system through a dynamically and procedurally generated simulation environment, based on real terrain and community data.

Keywords: Smart city logistic · Electric vehicle · Community study · Simulation

1 Introduction

Fully electrically powered vehicles help to get urban areas clean, silent and more attractive. The launch of the German national development plan for electro mobility [1] has spawned some activities and projects, ranging from research to industrial development, and sporting goals with short-term as well as long-term lifelines. In most cases the obvious shortcomings of currently available, fully electric vehicles (EVs) are addressed and tackled with a particular mix of various technologies. Unfortunately, the launch of EVs in Germany is challenging. Only 18.948 EVs were moving on German roads at the beginning of 2015 [2]. Concerning the plan of the German government about getting one million EVs at the end of 2020, a lot more of them have to be placed on our roads. Thus, the German BMWi funds multiple projects within the research program "Information and Communication Technologies for Electric Mobility II (ICT II)". This program is focusing on research about information and communication technology (ICT)-system architectures for EVs (Smart Car), EV integration in energy supply infrastructure (Smart Grid) and intelligent traffic infrastructure (Smart Traffic). One great

© Springer International Publishing AG 2016
G. Fahrnberger et al. (Eds.): I4CS 2016, CCIS 648, pp. 217–222, 2016.
DOI: 10.1007/978-3-319-49466-1_15

enabler within this domain could be the commercial usage of EVs in modern cities (Smart City). More than 60 % of newly registered vehicles in Germany belong to purely commercial used fleets in 2014 [3].

The Smart City Logistik (SCL) project [4] targets the application domain of inner city merchandise traffic. The concept is to unload cargo from heavy trucks on the cities perimeter and to run the last miles with small and medium sized EVs. In most cases the logistics partners also utilize storage facilities outside the city to provide additional buffer capabilities and to decouple long-range from short-range traffic in this way. The challenge is to support mixed fleets with EVs. This can be done with an ICT-system that provides interfaces to existing logistic systems, an intelligent route planning system with dynamic range prediction and supporting capabilities for dispatchers and drivers with real-time feedback, optimizing the driving style and providing alternatives for successful ends of planned tours.

Field tests and acceptance studies, as used in SCL, can be found in many other projects related to EVs. For example sMobility, a project about a cloud-based system- and service platform to combine street, EVs and grid operators, wants to evaluate their system within an acceptance test by using 200 EVs [5]. Unfortunately, just 76 vehicles were available. Another one, eTelematik, a project about a solution for municipality using a special prototype of Multicar as an EV with different extensions like a road sweeper or a snow plough, has to evaluate the ICT-system in combination with their newly developed telematics unit [6]. iZEUS, a project about complex standards for managed charging using decentral storage in EVs with an energetic recovery system for B2B car fleets, evaluates their system with 90 private and 30 business customers in combination with their provided 600 charging points [7]. Lastly, EWald, a collaborative research project with new developed and intelligent charging devices and communication concepts for EVs, evaluates their solutions within 7,000 km of rural area by using 230 EVs [8]. Each of them has one in common: testing the system solution requires a lot of experiments, mostly a lot more than available.

Thinking about ICT-system evaluation and the required amount of experiments, the question about how to support these studies with much more practicable approaches rises. How could these approaches reach more experiments without punishing the budget? The presented solution tries to support a community-based evaluation process (CBEP) by using a simulator, with highly procedurally generated landscapes based on real map data, which can be used in combination with the finally available SCL ICT-system, without modifying them.

2 CBEP Simulator

CBEP is a cabin based on a former multi-purpose vehicle Multicar, which is famous in city and pedestrian areas. The cabin is fully featured with all equipment for the everyday work in courier, express and parcel area but is not able to move. The cabin contains a simulation environment and a SCL ICT-system in which a driver can process several orders. So in the background the SCL system is working. In the foreground the driver interacts only with a driver assistant client (DAC) by driving in a simulated city generated from true roads with real-time position data.

Figure 1 presents the cabin schema. In front of the driver, a huge flat screen is the new virtual glass windshield. Besides the steering wheel, there are two additional small screens which form the cockpit. The cockpit touch screen allows switching cabin equipment like radio, air condition, lights et cetera on or off. On the second screen, a navigation system and the cabin state like velocity is accessible. The DAC is placed on top of the cockpit.

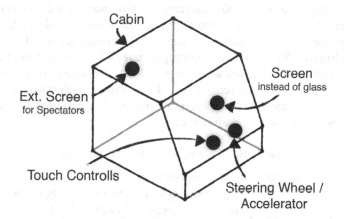

Fig. 1. CBEP cabin schema

Entering the cabin the considered test drive is an express order from a suburb to the city center delivering ice cream. Time for the order is around 5 min. In the beginning, there is an area to be familiar with the EV. All the time a virtual assistant driver is helping per voice. In the test drive, there is an average working day and no rush hour. On midway, the weather pattern is changing. At the end of the trip, the driver gets a report how to improve the next tour.

3   CBEP Data

Two cabin simulators were in parallel running around 6 months in European exhibitions. Test driver ages range from young to old. But all data sets have no personal items. Every 500 ms CBEP simulator's data logger writes a record of position, energy consumption and four additional parameters. Overall there are 10,822 trips with about 500 trip segments. So the entire data set contains 5,338,084 segments. Processing the data set all segments with velocity zero are removed, and trip segment's energy consumption is normalized by distance. So the distance parameter could be eliminated in the range estimation model. Therefore, it is necessary to remove all items with a distance less than 0.001. Otherwise, energy consumption scores absurd high. As a result, 5,097,588 segments remain. The next data processing step is to eliminate high performance shifts in negative acceleration and energy consumption. Such shifts are produced by high

Table 1. Relative error in parameter loading in percentage

Parameter	All	acc.	ac	Velocity	Weather
3 intervals	10.3	25.4	17.4	11.9	10.4
17 intervals	6.4	25.4	15.8	12.4	6.4

velocity crashes hitting a snag. This step is done by boxplot. That means for energy consumption items 89.4 % are inside the interval [−19, 3, 32, 1] and for acceleration items 90.1 % are inside the interval [−4, 5, 6, 9]. Thereby 564,198 segments are inapplicable, and a data set of 4,533,390 items remains. Now the adjusted data set is analyzed for parameter loading and range estimation model parameter mix dependency. Therefore, 200 trips are selected randomly. The trip segments range between 113 and 645. So on average a trip has 450 segments (overall trip average 420 segments). For the analysis, a Leave-One-Out cross validation with $k = n = 200$ was applied. Altogether, ten runs are processed split into two test series (3 and 17 intervals). Each run-series starts with all parameter for learn and test stage. Remaining runs are done by one parameter missing. Table 1 presents the parameter loading results.

Table 2. Relative error on average in percentage

Tour	Segments	Relative error on average			
		All param.	w/o acc.	w/o ac	w/o ac + velocity
10	4,586	9.08	29.31	19.28	27.44
20	7,970	8.42	26.52	18.83	25.32
50	19,657	7.56	26.92	18.56	25.04
100	42,087	7.87	26.86	16.40	24.47
200	81,443	7.49	26.27	15.80	23.97
500	205,909	7.66	26.92	15.22	24.01
1,000	418,344	7.52	26.61	15.26	24.07
2,000	830,729	7.63	26.95	15.35	24.17
5,000	2,086,953	7.51	26.40	15.26	23.99

That gets the following order of importance: (1) acceleration, (2) air conditioning, (3) velocity and (4) weather. For both runs, a missing weather parameter is not significant for epsilon and maximum relative error. In contrast, the remaining parameter (1)–(3) have a strong influence on the overall result. Hence, the weather parameter could be unattended for a better calculation performance. That is crucial for the second test series which provides good results. Unfortunately, the learning process takes 65 % longer than the first test series. So it is to deliberate about whether the increased computing time is worth it. Therefore, the quantity of the learning data set is crucial. To study the variation 10, 20, 50, 100,..., 5000 trips were selected randomly for learning data sets. Thereby, it should be noted that a data set is a trip mix. In a data set of 20 items, there are 10 trips from the previous set and 10 additional trips (for 50 items 20 previous and 30 additional trips and so on). Furthermore, there is no overlapping between learning and test data sets. So starting the test runs the model has to be constructed by a learning data set. Then the relative error was detected for each test run. Table 2 presents the relative error on average.

As a result, the range estimation model converges quickly even for small data sets. For the proposed test drive the relative error with all parameters remains around 7.5 %. Additional data sets only have a minor improvement for better quality. Without the loss of generality, the range estimation model enters the target corridor with 200 trips for any parameter combination. Furthermore, the range estimation model provides a reliable prediction of a learning data set of 200 trips. Figure 2 presents the energy consumption forecast during a drive based on a learning data set of 200 items. For eight tours with different routes, the figure shows the off between the prediction and the real energy consumption per trip segment. So for the whole trip the real energy tightly follows the predictive consumption (2(a)). But at tour starting prediction for single trip segments is really off (2(b)) which is not significant because at the starting point there only is minor total consumption. After a high fluctuation phase, the relative error converges straight to minor values.

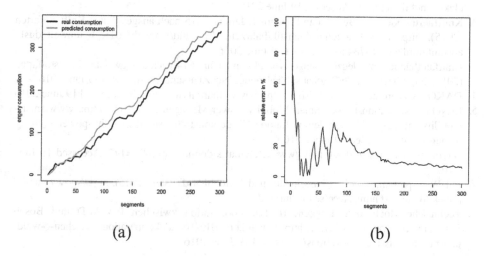

(a)   (b)

Fig. 2. Prediction vs real energy consumption

4 Conclusions

The SCL project has implemented and installed the system. Measure the correctness of system as well as the range prediction model was demonstrated by this community-based simulation environment and embedded consumptions behaves. Tests within the CBEP cabin produces data as required and can be modified to match any logistic scenario. The realistic visualization and handling within this cabin helps participants navigate those scenarios and proceed as normal. Those tests don't compensate the real test beds but simulated environment help to integrate a larger amount of participants than the test bed could yield. Furthermore data and results generated by simulation in statistic significant manner with real subjects were very helpful to find a way to deal with the challenges for improving the whole ICT-system. As a result the simulated system acts in similar way to the real system along with the behavior and correlated errors but the setting is much cheaper to reach statistic relevant results.

Acknowledgements. We would like to thank all members of the SCL research team here at FSU Jena – there are too many to name them all in person. We would also like to extend our gratitude to our partners within the research consortium, end users as well as research institutions and industrial developers. This project is supported by the German Federal Ministry for Economic Affairs and Energy in the IKT-II für Elektromobilität program under grant 01ME121(-33).

References

1. Bundesregierung Deutschland. NEP Elektromobilität (2012). http://www.bundesregierung.de/Content/DE/Infodienst/2012/10/2012-10-12-elektromobilitaet/2012-10-12-elektromobilitaet.html. Accessed 19 June 2016
2. Kraftfahrtbundesamt. Bestand an Pkw am 1. Januar 2015 nach ausgewählten Kraftstoffarten (2015). http://www.kba.de/DE/Statistik/Fahrzeuge/Bestand/Umwelt/2015/_b_umwelt_dusl_absolut.html?nn=1378446. Accessed 19 June 2016
3. Kraftfahrtbundesamt. Neuzulassungen von Pkw im Jahr 2014 nach ausgewählten Kraftstoffarten (2014). http://www.kba.de/DE/Statistik/Fahrzeuge/Neuzulassungen/. Accessed 19 June 2016
4. DAKO. SCL project website (2016). http://www.smartcitylogistik.de. Accessed 19 June 2016
5. INNOMAN. Demonstrator - Smart Mobility Power Management (2015). http://www.smart-mobility-thueringen.de/weg/demonstratoren/77-demonstrator-smart-mobility-power-management. Accessed 19 June 2016
6. Navimatix. Feldtest (2014). http://www.etelematik.de/index.php?id=172. Accessed 19 June 2016
7. EnBW. Erprobung von Ladeinfrastruktur und EMobilitätsdiensten (2015). http://www.izeus.de/projekt/flottentest.html. Accessed 19 June 2016
8. Technische Hochschule Deggendorf. Ladekooperation zwischen E-WALD und Bosch Software Innovations (2015). http://e-wald.eu/2015/10/ladekooperation-zwischen-e-wald-und-bosch-software-innovations/. Accessed 19 June 2016

Author Index

Aguiar, Hugo   153
Apel, Sebastian   217
Ayaida, Marwane   143

Barbosa, Ricardo   153
Bartnes, Maria   129
Boavida, Fernando   153

de Koning, Marco   49
Dias, Hugo   153

Ebner, Martin   3
Egas, Carlos   153
Eichler, Gerald   211
Erfurth, Christian   211

Fahrnberger, Günter   90
Fengler, Olga   115
Fengler, Wolfgang   115
Figueira, Ashley   153
Fouchal, Hacène   143

Ganishev, Vasilii   115
Gebhard, Kai   217
Gillard, Didier   39

Herbin, Michel   39
Herrera, Carlos   153
Hussenet, Laurent   39

Jaatun, Martin Gilje   129
Jeon, Hyung-Jin   59

Khalil, Mohammad   3

Lommatzsch, Andreas   173
Lüke, Karl-Heinz   211

Mauch, Marianne   217
Meiners, Jens   173
Merniz, Salah   143
Moussaoui, Boubakeur   143

Natvig, Marit K.   192
Nunes, David   153

Park, Joo-Hyuk   59
Pereira, Vasco   153
Phillipson, Frank   49

Raposo, Duarte   153
Reis, André   153
Rodrigues, André   153
Rossak, Wilhelm   217
Roth, Jörg   23

Sá Silva, Jorge   153
Schau, Volkmar   217
Sinche, Soraya   153
Son, Seoung-Woo   59

Tøndel, Inger Anne   129

Westhoff, Dirk   75
Wienhofen, Leendert W.M.   192

Yoon, Jeong-Ho   59

Zeiser, Maximilian   75